Forest H. Belt's

Easi-Guide to COLOR TV

HOWARD W. SAMS & CO., INC.
THE BOBBS-MERRILL CO., INC.
INDIANAPOLIS · KANSAS CITY · NEW YORK

FIRST EDITION

FIRST PRINTING—1973

Copyright © 1973 by Howard W. Sams & Co., Inc., Indianapolis, Indiana 46268. Printed in the United States of America.

All rights reserved. Reproduction or use, without express permission, of editorial or pictorial content, in any manner, is prohibited. No patent liability is assumed with respect to the use of the information contained herein.

International Standard Book Number: 0-672-20936-5
Library of Congress Catalog Card Number: 72-91726

Contents

CHAPTER 1

What Color Television Is (and Does) 7
Color television from camera through control room, transmitter, and tower, to the receiver. Three rasters and pictures go together and make one black-and-white before you can see true color.

CHAPTER 2

Knowing One Color Set From Another 19
Styles of color television cabinets, sizes of picture tubes, and features you're likely to see. Today's modular designs. How to build your own color television receiver.

CHAPTER 3

Words You Ought to Know 37
Learning the lingo of color brings advantages. You'll come nearer picking the best set for you and then enjoying it to the utmost.

CHAPTER 4

Buying Your Next (Or First) Color TV 47
Where to go shopping and what to take along. Compare several sets, both name-brand and others. Decide whether a particular model fits *you*. Win the warranty "game."

CHAPTER 5

Antennas Improve Color Reception 57
When and why you need an outdoor aerial, and when not. Antennas suit different localities. Prices. Indoor antennas. Cable television and you. Apartment antenna systems.

CHAPTER 6

Good Installation Avoids Problems 71
    When they deliver your new color tv. Uncrating. Learning how your set operates. Running antenna lead-ins the right (and wrong) way. Lightning protection. Adjustments that should be made as soon as the set is installed. Don't miss out on your new-set warranty.

CHAPTER 7

Tuning Your Very Best Color Picture 87
    The importance of fine tuning and how to get it just right. Setting Horizontal and Vertical Hold for the steadiest picture. Refining Contrast and Brightness for color viewing. Best settings for Color and Tint.

CHAPTER 8

Fixes You Can Try (and Some You Shouldn't) 105
    Troubles at the top and bottom, plus cures. Horizontal and vertical hold and centering. Screen impurity and how to "track" gray scale. Focus. High voltage. Ghosts and other picture annoyances.

CHAPTER 9

Trimming Color TV Repair Costs 135
    The secret is not always in doing it yourself. You may need a top-notch repair technician. Ways to find one and ways not to. What to do when you've found him.

Preface

Few inventions have changed modern home life so much as television. Almost everybody has a tv receiver. Many families have two or three. Television helps breakfast get the day started, relieves boredom in the afternoon, babysits youngsters after school, serves up evening news, and lulls away the hours till sleeptime. For good or bad, everyday living revolves around most tv sets.

More recently, fascination with television has switched to color. Upward of half the homes in the country boast color tv sets. All network television programing comes through in color. Most local stations telecast their own color. At school, a growing percentage of the television films and video tapes your kids study from are in color.

If you don't already own color television, the odds are that you'll shop for one before another year goes by. The old excuse that "color tv isn't good enough yet" just doesn't cut it anymore. Go somewhere and watch a dependable brand of color set that has been properly installed and adjusted, and you'll at least question that old notion. Prices are still higher than for black-and-white sets, and that won't change. But you can buy a color tv receiver for easily under $300. So cost won't likely hold you back much longer, if that's why you're waiting.

Maybe you watch color already. If you have a good model and feed it plenty of tv signal from a strong outdoor antenna, you know color tv nowadays can look great. You've seen the beauty most programs offer in color. Certain special shows *depend* on color for special effects: psychedelic decorations, turned-on illumination schemes, way-out flashes of light and color. Once you view much color, black-and-white tv appears almost dull.

Whether you own a color tv or plan to shop for one, you need very much to know three things about color.

(1) *How to get the best color picture.* Some of the knobs are strange, compared to black-and-white, and tuning in the best color has its own special tricks. The way your color set is installed and adjusted determines how good a picture you'll get (and for how long). You need to know what's right.

(2) *How to save money on repairs.* Color tv sets are complicated. Many use transistors, which increase dependability but preclude "home" repairs. You should know what's safe to try yourself and what's not. And when you need a good repair technician, there *are* ways to find one.

(3) *Where and how to buy.* If you are in the market for a color set, you have matters to consider like: size, appearance, quality, price, warranty, delivery, installation, and even future service. Satisfaction from your first color tv, or from your third or tenth, depends on what buying decisions you make.

This little book is packed with answers to questions asked most about color tv. Soak up these pictures and words from cover to cover before you head out to buy a color set. Again, before they deliver a new one, browse the chapters on antennas, on installations, and on tuning in a top-quality picture.

Afterward, keep this book where you can find it. You'll find the photos and explanations vital when you run into trouble with any color tv, new or old.

Enjoy color. Keep yours the best. Television *is* better in color.

Forest H. Belt

Chapter 1

What Color Television Is (and Does)

Color television began getting into gear around the middle of the 1950s. But for ten years, progress wasn't exactly swift. Of the three television networks, only NBC really pushed color in those years—and NBC did mainly because its parent (RCA) devised our present means to transmit and receive color.

By the late 1960s, CBS and ABC had joined NBC in telecasting almost exclusively in color. Eighty percent of local stations had color studios. Today, all network shows are in color. Millions of viewers take color for granted. Cartoons or commercials, sports or news, quiz show or soap opera. Almost everyone can have their television viewing in color. The photos that follow tell how.

HERE'S WHERE IT ALL BEGINS. A color telecast originates from a special three-color tv camera. Everything you and I see reflects light that corresponds to certain mixtures of three primary colors: red, green, and blue. Orange, for example, consists of red and green, with emphasis on red. Purple comes from blue and red, with mostly blue. Brown mixes unequal proportions of all three.

White develops when exactly the right amount of all three basic colors mix. You see black when there's absolutely none of any color. Gray is small amounts of all three primary colors, but mixed in exactly the same proportions as for white.

The color television camera has three color filters. From whatever scene the camera "sees," one filter lets through only whatever amounts of blue exist in the scene. Another filter passes only red, and the third, only green. Electronic circuits inside the camera turn these various amounts of red, green, and blue into electrical impulses called red, green, and blue *signals*.

IN A COLOR TELEVISION CONTROL ROOM, television directors and producers switch from camera to camera to show different views in the studio (or on the golf course or ball field). They can switch on film cameras that convert colored news film or commercials to electrical signals. Projector drums for color slides work the same way.

Cameras, projectors (film "chains," they're called), and slide drums each produce three signals that correspond to the three primary colors their filters separate out of a scene. In the racks of electronic circuitry around the control room, the three primary signals are specially processed and mixed with other signals to make a total color tv signal. The colors are called *chroma*. Other parts of this total signal are *video* and *sync* (short for synchronization pulses).

This video/chroma/sync signal has everything your color tv receiver needs to make a color picture. But these signals must get to your tv set some way. That's the job of a station transmitter.

THE TRANSMITTING TOWER is usually all you see of a color television station. It stands on some high hill, usually outside of town, or on a tall building. Here, the transmitting antenna sits atop a tall tower.

The domes on the side of this tower are *microwave* antennas. The video/chroma/sync signals and an audio (sound) signal are beamed out to the transmitter site by microwave from the building that houses cameras and control room. Equipment in the transmitter room (inset) takes the signals received by the microwave antennas, mixes them with the station's channel signal, and transmits them into the antenna, which radiates the whole signal into the air.

Take a station on channel 8, for example. The channel-8 station signal is created right here in this transmitter room. To it is added the video/chroma/sync and audio from the studio control room, by a process called *modulation*. That modulated channel-8 signal is what your color television receiver picks up whenever you set the channel switch to 8. Circuits in the receiver *demodulate* the video/chroma/sync and audio signals, and turn them into a color picture and program sound.

THE SCREEN OF A COLOR TELEVISION PICTURE TUBE looks like this if you view it close up through a magnifying glass. The little round dots are arranged in triangular groupings. In each group or *triad,* one dot is red phosphor, one green phosphor, and one blue. They light up whenever electron beams inside the picture tube "shine" on them.

The phosphor dots are so tiny that your eye blends the lighted colors. That's why a color picture tube can reproduce for you all the colors the camera sees originally. The red chroma signal, demodulated and processed by the tv receiver, paints the red portions of the televised scene, in exactly the same places and in the same proportions the red camera filters saw. The green chroma signal lights up certain green phosphor dots and re-creates whatever green exists in the original scene. Likewise, the blue signal turns on blue dots, corresponding to whatever blues or blue mixtures appear in the scene at the studio (or ball park, or in a film).

If the original scene is televised in black-and-white, as with an old, old movie or with some educational tv programs and by a few small stations, phosphors for all three colors light up in exactly the proportion to produce white and shades of gray.

ITS CONTROLS DISTINGUISH A COLOR TV receiver. Two knobs in particular identify a color set. One is labeled Color, the other usually labeled Tint. Generally they're together on the front panel.

The first affects how brilliant the colors are that you see on the screen during a color telecast. The other, the Tint knob, varies the hue of colors you see. In fact, in some models the knob is labeled Hue.

Just above the knobs on the color receiver in this photo is another identifier. The printed color label and the monogram are typical. Almost all color sets have a two-color or three-color monogram plainly visible somewhere.

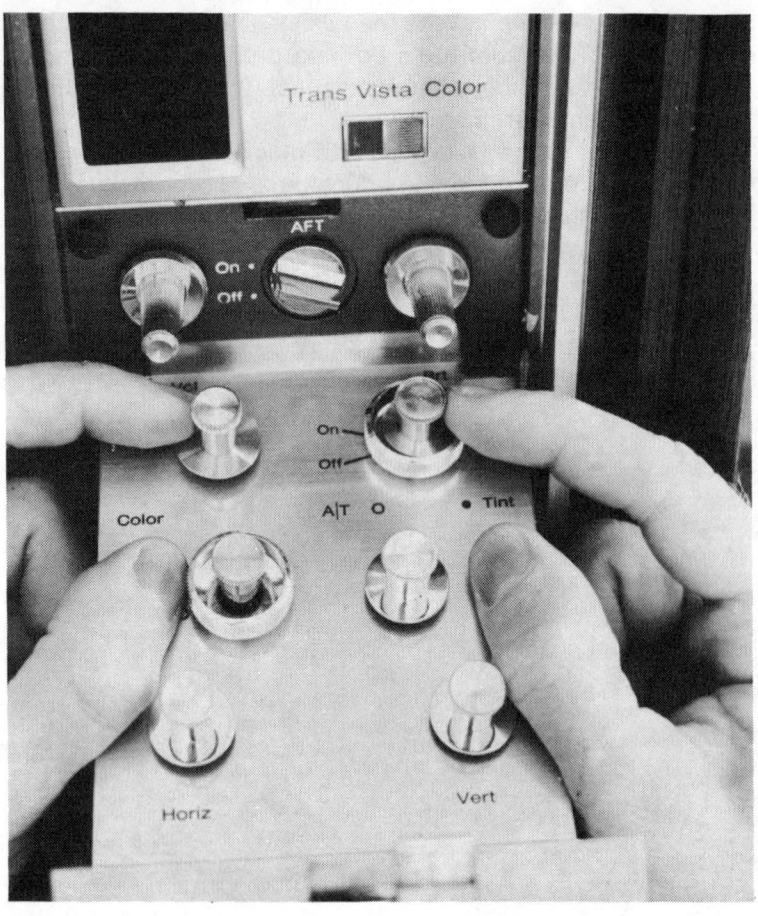

12

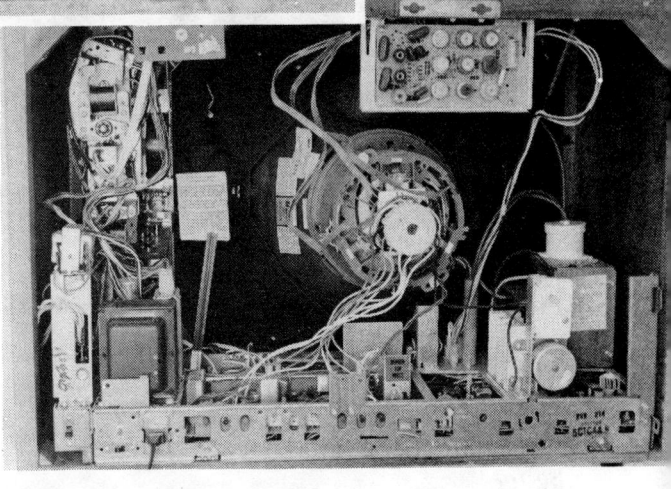

YOU CAN SEE MAJOR DIFFERENCES between color and black-and-white with the back cover removed. For one thing, that color picture tube is bulkier for the same screen size. Compare it with the black-and-white chassis in the inset.

The *deflection yoke* is the large gray housing around the neck of the picture tube. It magnetically sweeps the red, green, and blue beams back and forth on the screen. You might relate the yoke's action to flipping a flashlight beam back and forth, except in the color picture tube there are three beams. The yoke in a color set is large and has more wires going to it than in a black-and-white tv. The black-and-white yoke (inset) sweeps only one beam.

Color-tv innards are larger than for black-and-white. Always they're heavier. And, if you notice along the back of the chassis, you'll see the color set has more adjustments.

13

A CLOSER LOOK AT NECKS shows how different a black-and-white picture tube (inset) is from a color version. Most significant is the large structure on the neck of the color picture tube. (A picture tube is often called "crt" which stands for *cathode-ray tube*.)

The big white structure you see on the color crt houses an assembly called the *convergence yoke.* Notice the small knobs and thumbwheels on the white covering. This convergence assembly assures that each beam inside the color crt hits its color of dot (on the screen) accurately.

The deflection yoke, you recall from the preceding page, sweeps the three color beams back and forth across the screen. It simultaneously, but more slowly, pulls the beams from top to bottom, so the beams light up all the phosphor dots over the whole screen. The convergence yoke helps the deflection yoke sweep the three beams exactly the right speed to coincide with sweep timing at the tv station and camera. Thus the red, green, and blue dots on the receiver screen light up in sequences that exactly match the colors picked up by the color television camera.

WHAT APPEARS ON THE SCREEN of a color television receiver when you have no station tuned in is called the *raster*. If the deflection yoke is moving all three beams normally, and the convergence yoke has the timing precise, the raster you see should be white (or nearly so). It should look just like a black-and-white receiver looks with no station.

But remember that the color picture tube must mix red, green, and blue in just the right proportions to make white. If, for example, there's too little green and blue, the raster looks pinkish. Too little green leaves the raster purplish or magenta. Too little blue gives the raster a yellow cast; too much green turns it greenish.

The Screen controls illustrated here set how much of each color goes into the raster. You can think of the color screen as showing three individual rasters superimposed. Starting with all three Screen controls at zero, the screen is black. Turning up red gives a red raster. (Green alone would make a green raster, or blue a blue.) Adding green to red brings a yellow raster. Adding blue turns the raster white. But only if all three colors are in exactly the correct ratios.

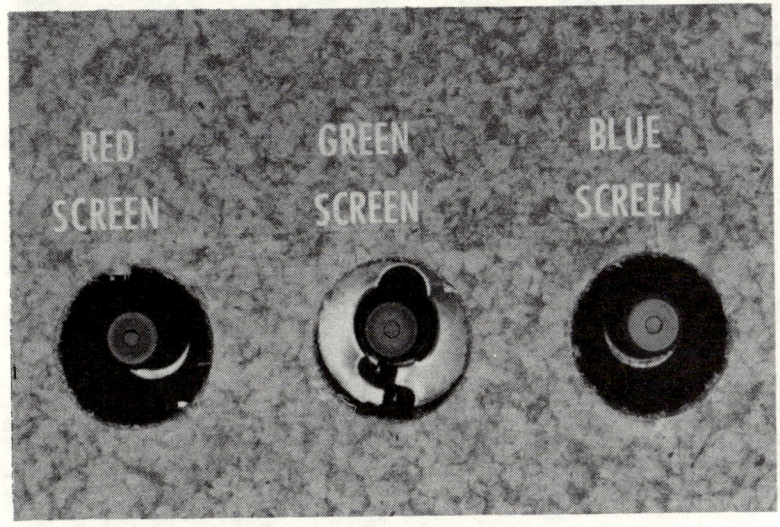

15

A BLACK-AND-WHITE PICTURE ON A COLOR SCREEN has the same three-color makeup just described for a raster without any picture. There are, you might say, three separate pictures. In this photograph, each one has been moved in a different direction to split them apart for you to see. One is red; one, green; the other, blue.

As with a blank raster, the trick is for the picture tube to mix exactly the right amounts of red, green, and blue to make shades between white and black show up as true gray. That job belongs to three Drive controls (inset). Some color sets have only blue and green Drive knobs. A knowledgeable technician adjusts them and the Screen knobs carefully to give perfect black-and-white rendition when there's no color in the picture.

AND THAT'S IMPORTANT. Keep this in mind. No color receiver can show true colors unless the Screen and Drive controls mix just the right proportions of red, green, and blue to make a *white* raster and an *unshaded* black-and-white picture *when there's no color.* (You can check this with any program by turning the Color knob down.)

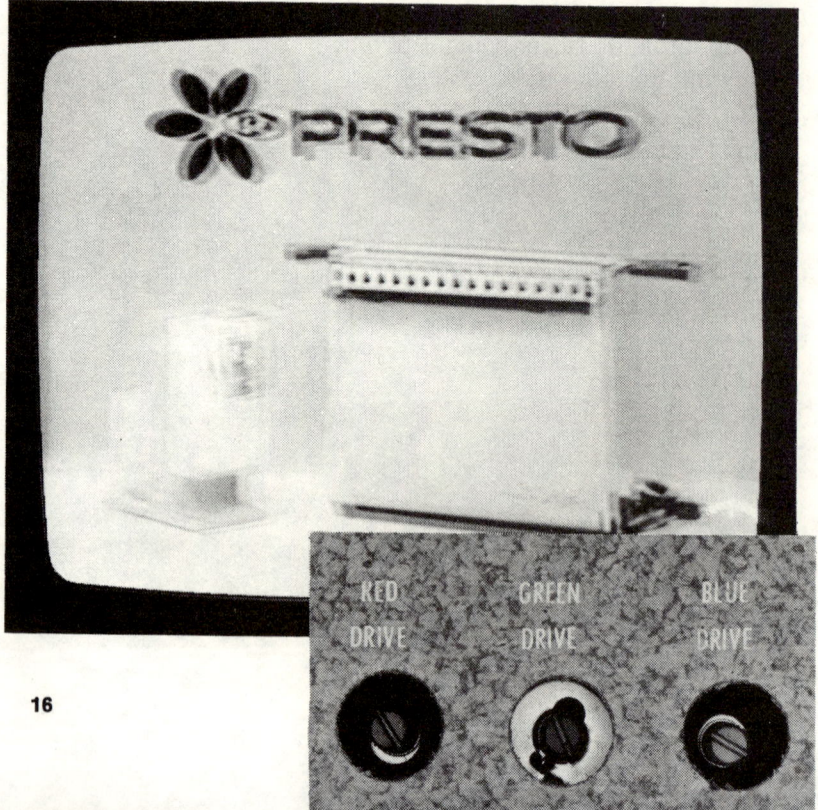

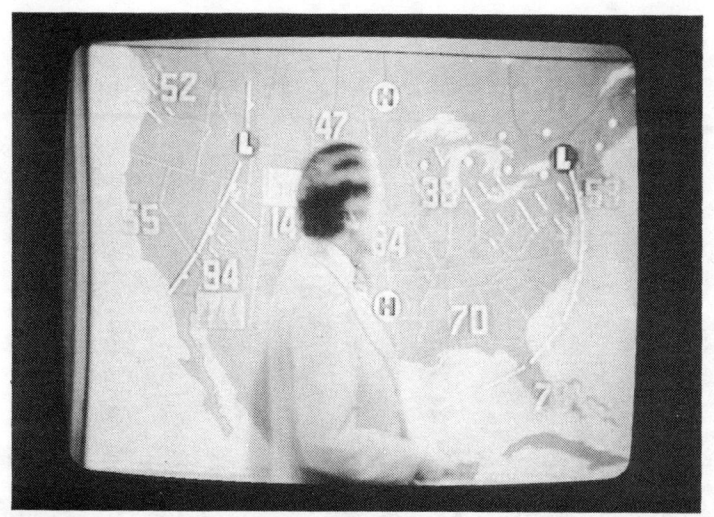

A NORMAL COLOR TELEVISION PICTURE results, then, from all the things you've seen in this chapter. A color tv camera (or film chain) sends video/chroma/sync and audio signals by microwave to a transmitter and antenna (the latter is on top of a tall structure—usually a tower). Your color tv antenna (Chapter 5) picks up the station signal, and your color receiver recovers the video/chroma/sync and audio. Circuits in the receiver turn the audio into sound and the video/chroma/sync into a color picture.

Before the receiver can "paint" an accurate color picture, the crt screen must be able to show a pure white raster. Then, with no chroma (color), the video, too, must show a black-and-white picture—not three (opposite page), but one picture image as you see above. The color (red, green, blue) content of those three images is determined—as long as there's no chroma—by the Drive controls.

The *positioning* of those three images depends on that convergence yoke mentioned earlier. The deflection yoke paints the raster, but the convergence assembly makes each of the three beams sweep precisely. On the opposite page, convergence is messed up. Above, convergence knobs have been turned to superimpose the three pictures exactly.

Now, with Color turned up, the set can show an accurate color picture.

17

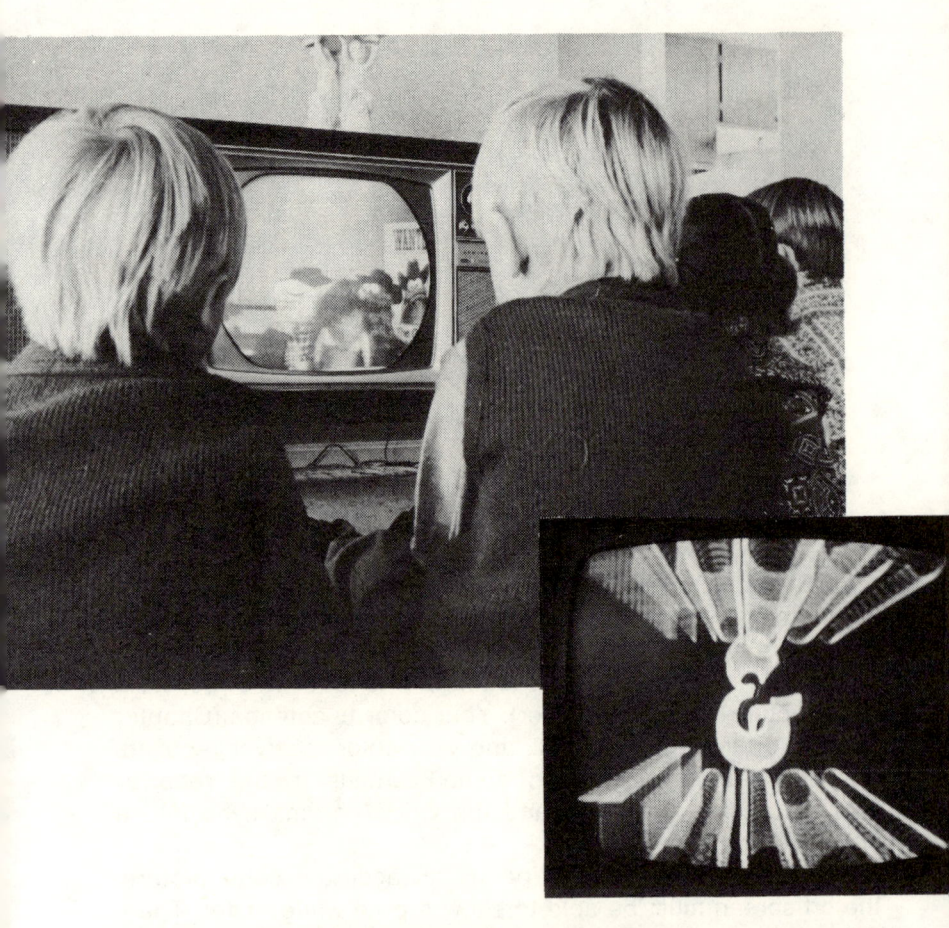

COLOR TELEVISION PROGRAMS like *Sesame Street* and a new art series called *Images & Things* lose a lot of their impact in black-and-white. That's why schools all over the country that use educational television pinch their budgets to replace outmoded black-and-white sets with new color models. At home, even on old color receivers, well-designed and produced color programs hold youngsters at rapt attention.

Until you've seen an Apollo space shot in color, or sat glued through a true-to-life *National Geographic* travelogue, or watched a World Series on tv-green turf, or even watched *The Flintstones* romp through their cartoon-color escapades—you just haven't seen television the way it can be. Once you do, chances are good you'll turn on to color, too.

Chapter 2

Knowing One Color Set From Another

At one recent count, 176 different color tv chassis (the insides) were available. More than 1000 different models (cabinet/chassis varieties) could be bought in the United States. Some 30-odd different companies either build or import color television.
And that's just for one model year. Each year, several more models appear. Some replace earlier versions, some add new conveniences to old frames, some are the same chassis with new numbers. The proliferation makes picking out which set to buy something of a problem. The list of possible cabinet-and-chassis combinations from just one major set-maker scatters confusion among salesmen and customers alike. Imagine what it's like sorting out all the competitors.
Best you know before you start what's available. Sizes go by diagonal measurement of the viewing portion of the picture tube. Sizes range from a tiny 4-inch to the giant 25-inch picture tubes with square corners. In between, you find 9, 10, 11, 14, 16, 17 (another square-corner type), square-corner and regular 19-inchers, 21-inch square, 22-inch, and the old 23-inch (what once upon a time was called 25-inch). Picture tubes with a new kind of screen known as *black-surround* deliver brighter pictures. Sharply square-cornered crt's, with almost flat faces, fit more precisely the normal 3 by 4 dimensions of television pictures.

You have a choice among chassis that use all tubes, those that include a good many transistors, and a solid-state variety in which the only tube is the crt. Transistors lend an aura of advanced design, yet in some brands they offer little improvement over tubes. Solid-state chassis cost more, but in some designs greater dependability justifies the extra cost.

A recent innovation promises cheaper upkeep: *modules*. Parts are arranged on individual subassemblies that can be easily uplugged for replacement. The chief benefit seems to point toward less expensive repairs.

Cabinet designs illustrated in this chapter are fairly standard among set-makers. Special woods can add several hundred dollars to the price of a color receiver. You have to decide whether your color-tv budget permits fancy furniture or bare utility.

Read over the next pages. Study the photos. When you walk into a color-tv store, you'll know more than most walk-ins about the sets the salesman has on display.

SMALL-SCREEN COLOR TELEVISION SETS went through a period of high popularity, and some sizes still sell reasonably well. Small portables were almost the sole province of Japanese imports, until General Electric introduced a 10-inch set called the "Portacolor." The top photo here shows a 4-inch Panasonic; the bottom, a 9-inch Sony.

THIS LARGE-SCREEN COLOR RECEIVER HAS DANISH STYLING. The wood is walnut.

One factor in deciding fair prices for the furniture (the cabinet) of a color set is the finish. In some expensive woods, only veneer is available. Ask someone familiar with woods to advise you. When you spend a hundred dollars or more for fancy cabinetry, you should be sure of what you're getting.

SIXTEEN-INCH COLOR TV SIZES fall into two cabinet styles, both illustrated here. The Zenith 16-inch on top comes classed as a portable. It has a handle. This is the largest screen size categorized as a portable. You plug it into the power line, like all other color sets. You can't run it from batteries. But the cabinet is "light" enough to carry from room to room, if you've got the muscle for it. This particular model even has remote control.

The lower picture displays a table model in the 16-inch size. The wood cabinet is pecan, and styling is contemporary. You usually pay more for a table model than for a set classed as portable, but the cabinet looks more like furniture.

The 16-inch is about the smallest size still called a table model. A few 18-inch sets have handles, but they're not easy to move around. Notice, too, neither of these 16-inch models has the new flat-face square-corner picture tube (it's not available in that size).

NINETEEN-INCH TABLE MODEL color set is one of a line being marketed by RCA. It boasts several innovations.

For one thing, the picture tube is square-cornered and flat-faced. More than that, however, it was the first color tv to utilize wide-angle deflection. Most color television receivers have a deflection yoke that sweeps the beams over an angle of 90 degrees to scan from one side to the other. This picture tube uses a 110-degree deflection yoke. This picture tube neck is 4 inches shorter than usual. That means the cabinet can be shallower than regular 19-inch cabinets.

Another major feature is modular construction, already mentioned briefly. More than half the circuitry mounts on plug-in replaceable modules. Some of them are a special new kind; they're called *ceramic* modules. More about them on page 34.

THIS ALL-TRANSISTOR 19-INCH COLOR TV by a comparatively little known importer, Mitsubishi International, goes under the brand name MGA. The company is a relative newcomer to color television in the U.S., but it broke in with something of a splash.

Besides its 19-inch square picture tube, the set has some special features. One: a 25-position uhf tuner that lets you pick out any 25 uhf channels (more than you have in any one locality). Another: preset Color and Tint controls that can be adjusted to look best on one favorite channel and activated by the flick of a button.

Most important, though, and what attracted the most attention for this set: the chassis slides out and unfolds for servicing (see page 32). This doesn't make a better picture, nor does it help the set work any more dependably than another. But when there is trouble, the technician you call for service can do a faster job than with many other transistor color tv chassis.

25

THESE BIG-SCREEN ZENITH COLOR RECEIVERS illustrate Mediterranean styling (above) and French Provincial (below). Both these 25-inch sets include solid-state (all-transistor) chassis, with some of the circuitry on special modules (see page 35). Zenith chassis designs use several tiny devices called *integrated circuits* (ICs). Both these models have remote control.

TRINITRON PICTURE TUBE makes something special out of this 17-inch Sony table model. A handle lets you carry the set, but it's not the lightest thing you ever tried to lug.

The *Trinitron* picture tube is something entirely different from the triad-dot crt used in other color sets. Sony patented this color crt and is the only manufacturer using it. The three phosphor colors are deposited on the screen in vertical stripes rather than in dots. There's no need here to go into the technical implications, but the scheme does simplify certain circuits in the set. It also makes convergence—superimposing those three pictures perfectly—easier than with other brands.

Sony claims higher brilliance for the Trinitron, but that's hardly noticeable compared with the new "black-surround" picture tubes now being used in American color sets. This receiver, like all Sony models, is all-transistor. And they all use the unique Trinitron picture tube.

REMOTE CONTROL is one of the features you'll find handy in higher-priced color receivers. This model, one of the more elaborate, controls both vhf and uhf tuning, sets loudness of sound, varies color strength on the screen, moves color tint in either direction to derive perfect flesh color, and turns the set on and off. All this is done silently, except for what noise the channel-tuning motors make.

Some remote control systems affect only volume, on-off, and channel tuning. Several have motors on the volume control, color control, and so on. The newest systems use no motors. Instead, they have what are called *memory modules.* Special circuits "store" the setting you want and alter the control circuits electronically instead of mechanically. Result: silent operation of the remote system.

A small hand-held transmitter sends out inaudible vibrations. The tones are produced electronically by this transmitter, but mechanically by a few. See the tiny microphone in the grille between the two channel-selector knobs? That picks up the vibrations and sends them to the remote-control circuits and memory modules.

28

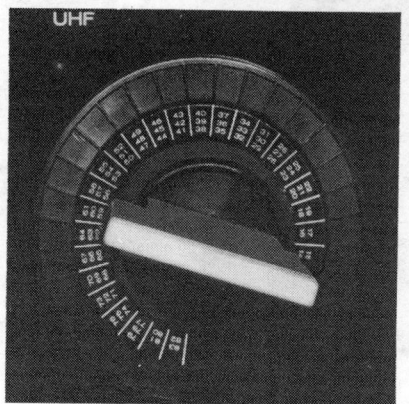

ULTRAHIGH-FREQUENCY (UHF) TUNING makes a considerable difference among new color sets. The Federal Communications Commission has ruled that eventually all new television receivers (black-and-white, too) must tune as easily to uhf channels as to vhf. Up till the ruling, continuous tuning (like with the 14–83 round dial) has been the commonplace. That's touchy. Although it works fine for black-and-white, continuous tuning is overly sensitive for color television—which tunes critically anyway.

On the way to complying with the FCC order, some manufacturers have come up with 25-position tuners. The one shown here is on an MGA receiver (page 25). Each position can tune one of three channels. You adjust, when you get the set, to whatever channels are active in your town. Then, snap-snap-snap, you can flip to uhf channels just as you do to vhf channels.

The third photo shows a hybrid. The little slide-rule dial is typical of many used throughout the years in uhf tuners. This one is different in that the knob has six snap (called *detented*) positions for whatever uhf channels are active locally.

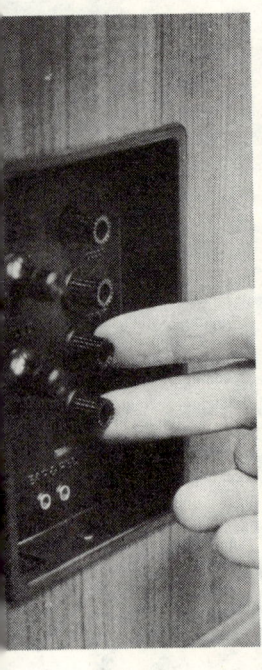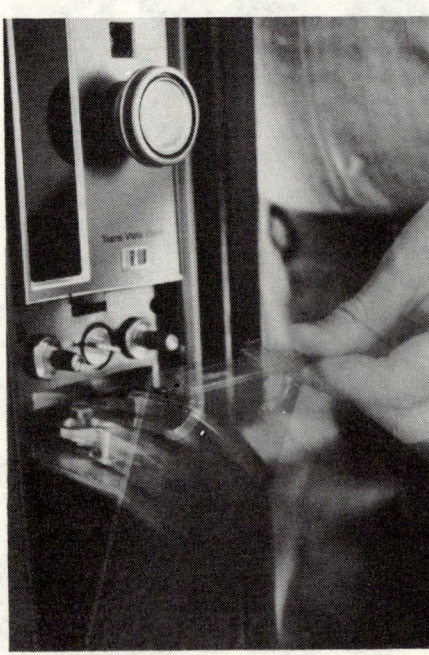

WHERE THE CONTROLS ARE makes a difference in how a color set looks and how easy it is to operate. Some controls may be inside a recessed panel called a *soapdish*. That name began years ago when such panels and their covers did resemble a soapdish. Some still do.

Some of today's soapdishes are on the side, without covers, as the left-hand photo above illustrates. This model has brightness, contrast, vertical and horizontal hold, plus color and tint knobs, all in the soapdish. To boot, little white jacks near the bottom let you plug in earphones.

A trapdoor hides the control knobs of the color receiver in the center photo. Two holes give access to volume control and brightness knobs when the lid is closed. The remaining controls are hidden from view except when you want to adjust them. You have to watch walking past a protrusion like this if it's inadvertently left open; it snags.

At right above, vertical hold, contrast, and brightness knobs are almost completely hidden under the front edge of the cabinet. Even on a tabletop, there's enough room to manipulate these "thumbwheel" controls. Color and tint knobs are up by the channel selector, across from the on-off-volume control.

BLACK-SURROUND COLOR PICTURE TUBES have made a considerable impact on the industry. They go by various names. "Black matrix" is one term applied by several manufacturers. Zenith, for example, calls its version "Chromacolor."

Black surround means the area of the screen around each phosphor dot is black. This has two effects. For one, no halo effect surrounds the dots when the beams strike them. Beam energy concentrates in the dots themselves, and they are consequently clearer to see individually.

The other effect relates to room light that reflects from the phosphor-dot surfaces, and particularly from the areas that surround each dot. The black surround stops such reflection. You gain the benefit of stronger contrast between the dots and their surroundings.

The photo at left on this page shows a magnified view of a lighted phosphor without the black surround. Dots are not as clear, and the effect overall is darker. Several feet away from this phosphor, you see only average brightness. The right-hand photo shows how sharply the lighted dots stand out on a picture tube that has the black surround. From normal viewing distance, the raster seems brighter. With a picture on the screen, contrast is sharper and deeper between black and white elements. Colors in a picture have more sparkle and brilliance. (When you're shopping, ask a salesman to show you the difference side by side.)

MAKING COLOR TELEVISION EASY TO REPAIR has become a serious concern for some tv manufacturers in the United States. Color tv chassis and their inner workings are extremely complicated at best; it takes hundreds of parts and thousands of connections to make an instrument pick up video/chroma/sync and audio signals and turn them into a color picture and sound. For years, little attention was paid to how easy or how hard a receiver was to service. Manufacturers let the technician worry about that.

Your service technician has to charge for his time and knowledge. With labor costs everywhere skyrocketing, repair costs have climbed, too. The only hope for relief was to make sets easier to open up, easier to take apart, easier to test, and easier to fix once the trouble is found. No set is anywhere near perfect yet, but definite advances have been made in a few.

The rear view on this page illustrates the high accessibility of parts and connections in this foreign-made MGA. This 19-inch solid-state color receiver is the first imported model to be designed for easy servicing. The following three pages describe innovations turning up among popular domestic brands.

MOTOROLA WAS THE FORERUNNER in this country of easy-access color television. As long ago as 1967, this company introduced its "Quasar" line of all-transistor color sets. Some manufacturers predicted that this unique line of color sets would not be accepted, but they were wrong. The "Quasar" pushed Motorola forward in color tv sales.

Three startling innovations began with the "Quasar." (1) It was the first all-transistor color set sold by a U.S. manufacturer. (2) The "works in a drawer" concept lets your service technician slide the whole chassis right out of the cabinet, from the front. Servicing and repair adjustments are quick to reach.

(3) The Motorola "Quasar" was the first color receiver to be built largely on replaceable modules (inset). A technician just unplugs and replaces a faulty module. He can get new modules or rebuild old ones later, at the shop. Some troubles, of course, are on the main chassis and need shop repair.

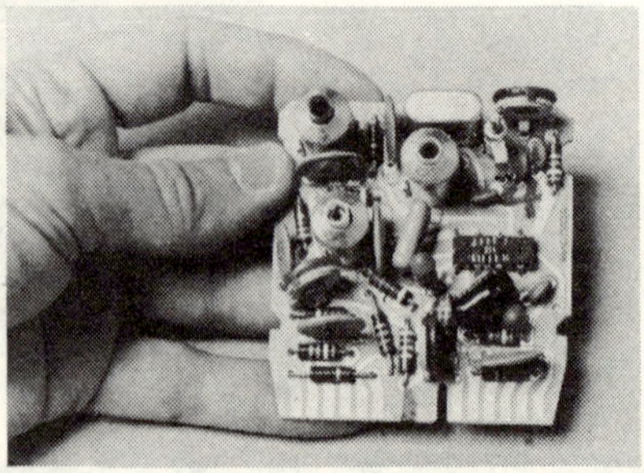

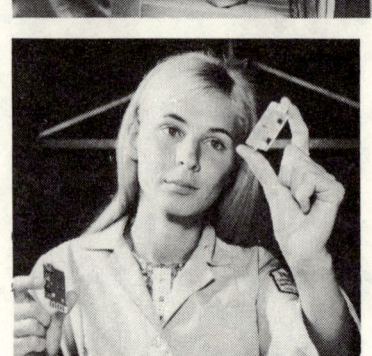

RCA DEVELOPED CERAMIC CIRCUIT MODULES. The attractiveness of building color television sets in replaceable subassemblies led RCA to its own version of modules. Conventional modules consists of flat fiberboards with metallic circuit wiring plated (or "printed") as foil strips on one side. Small parts—resistors, capacitors, coils, transistors, and diodes—are mounted with their wires through holes and soldered to the circuit foils. Some of RCA's modules are constructed this way.

But one RCA plant devised a flat ceramic material strong enough and good enough as an insulator to allow many of those small components to be printed or "grown" right on the ceramic base. Few other parts are needed.

One 18-inch RCA portable color model uses regular modules and a few of the new ceramic type. The technician in the photo is unplugging one (shown close-up in the inset) to replace it. Printed-board modules can be repaired. The ceramic kind are throwaway.

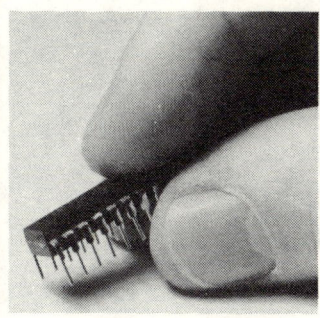

ZENITH CALLS ITS BOARDS DURAMODULES. They plug into the main chassis. Although seemingly only a small amount of the Zenith color circuitry, five modules actually contain a significant percentage of critical circuits. So much goes on the modules because of Zenith's use of integrated circuits.

Zenith puts virtually the whole chroma-processing section of its solid-state color television into integrated circuits. An integrated circuit is a tiny device (at right) in which resistors, capacitors, transistors, and diodes are formed by an etching process. You'd have to compare a regular color section, with its dozens of regular parts, to get an inkling of the space and wiring saved by an integrated circuit (IC).

35

BUILD A COLOR TELEVISION YOURSELF from a kit, if you want to really learn what goes on inside one. Heath Company, Benton Harbor, MI 49022, offers several different color models in kit form. You don't have to be any kind of genius to put one together, either; but you should be handy with tools and soldering iron. Step-by-step instruction booklets show you how to wire up each subassembly and then put them all together to form the all-transistor 25-inch color instrument you see here. You'll spend a lot of evenings, but seldom on so worthwhile a project.

These kits are not in the low-budget color tv class. Price is hefty. But the color set you build is not a skimpy design, either. Solid-state, with a varactor tuner (that's really the latest—see page 45), 25-inch square-corner black-surround crt, and a variety of cabinet styles, this is probably one of the best buys in color tv. You might even learn enough to service it yourself; Heath furnishes a small meter with the set, to help.

The cabinet shown is one of three styles available. Or, you can custom mount the finished kit in the wall, on a shelf, or in your own cabinet design. If you follow instructions carefully, you can have a highly satisfactory color television set this way.

Chapter 3

Words You Ought to Know

Wandering into the world of color television, you soon stumble over a pile of new words. Talk to people who know about color, and you hear them toss off words like "chroma" and "tint" just as casually as you mention "car" or "overcoat." Go to a store where color tv is sold and you get hit with "afc," "automatic tint," "varactor," "preset color," and so on. To hold your own, you need to know what they're talking about.

The pages in this chapter explain a few not-so-common terms. The pictures not only illustrate words that may be strange to you, but further your recognition of how a color television works.

Don't let the jargon bother you. You'll soon speak the language as well as the next fellow. Maybe better.

COLOR, CHROMINANCE, CHROMA

Every television picture in color consists of two signal elements. One is the video (explained on facing page) and the other is the color element. If they were separated, as they have been for the photo above, you'd see one picture that contains colors and another—just a fraction to the left—in black-and-white. Normally, the two pictures exactly coincide.

These two pictures, one called video and the other called *chrominance* or just *chroma,* even come through the color tv set separately. The circuits and stages that handle color are called the chroma section. Any trouble in your color receiver that affects color is called a chroma problem. In a few old color sets, the Color knob, which turns up the brilliance of color on the tv screen, was labeled Chroma.

It's enough to remember that the word *chroma* (*chrominance* is the formal version) refers only to the color portion of the tv signal or to circuits that process the color signals.

VIDEO, LUMINANCE, MONOCHROME

The other element of any color television picture is *video*. You might call video the contrast portion of the picture. Video determines how much or how little light goes into each segment of a televised scene. The Contrast knob lets you set the video to suit your own viewing preference.

A picture televised without any color is pure video. It's called a *monochrome* or black-and-white picture, and comes from the transmitter as a monochrome or black-and-white signal. No color or chroma comes with it.

A color television signal contains both video and chroma. A monochrome tv set processes only the video, so you see an ordinary black-and-white picture. A color set knows what to do with the chroma signal. The chroma signal and the video signal (often called *luminance* signal in color sets) make their ways through the set separately, but are brought back together at or near the picture tube. Chroma works its way through more slowly, so video must be delayed the same amount by a device called a *delay line*. The photos on these two pages show the signal with the delay line bad; you can see the chrominance and the luminance pictures separately. Normally, they superimpose precisely.

TINT, HUE, COLOR PHASE

The words *tint* and *hue* mean the same thing in color television parlance—at least most of the time. The Tint label applied to a knob on most color sets means that's what you adjust to make face colors right. A few tv makers label the same control Hue, on the theory that you alter hues when you turn the control. They are right, so either label is okay.

Your tv technician may call the control *color phase.* He knows that turning the Tint or Hue knob varies the phase (timing) of chroma signals demodulated from the station signal. Changing chroma phase makes the colors look different on the tv picture tube.

You can set Tint or Hue easiest while looking at closeups of an actor's face. Turn the knob all the way in one direction, you get a green face. Shifting phase to the other extreme makes flesh tones purple or deep blue. Somewhere between makes the facial hue that best suits you.

Hue alters from station to station and program to program. You may find yourself turning the knob several times throughout an evening, to keep faces looking normal.

AUTOMATIC TINT CONTROL, ATC

A long-time aggravation with color viewing comes from those uncorrected changes in color phase which occur at the station or on the network. So far, television broadcasting engineers haven't found a solution to this annoyance. Neither have receiver designers, for that matter, but a set of circuits called *automatic tint control* (abbreviated *atc*) help some.

You may hear someone mention "blue face syndrome" or "green face syndrome." The terms describe how flesh colors (which are close to a very pale orange) shift toward green or toward blue when you change stations, or another program comes on, or a commercial. To correct the hue, you ordinarily jump up and twist the Tint knob slightly (see opposite page). Many color-set manufacturers have recently included an automatic tint control (atc) to save you from so much knob-twiddling. The atc keeps flesh tones about normal in spite of chroma phase variations from the station.

Manufacturers have different trade names for atc and different labels for the controls that turn it on. The one shown here is called "Accutint" by RCA.

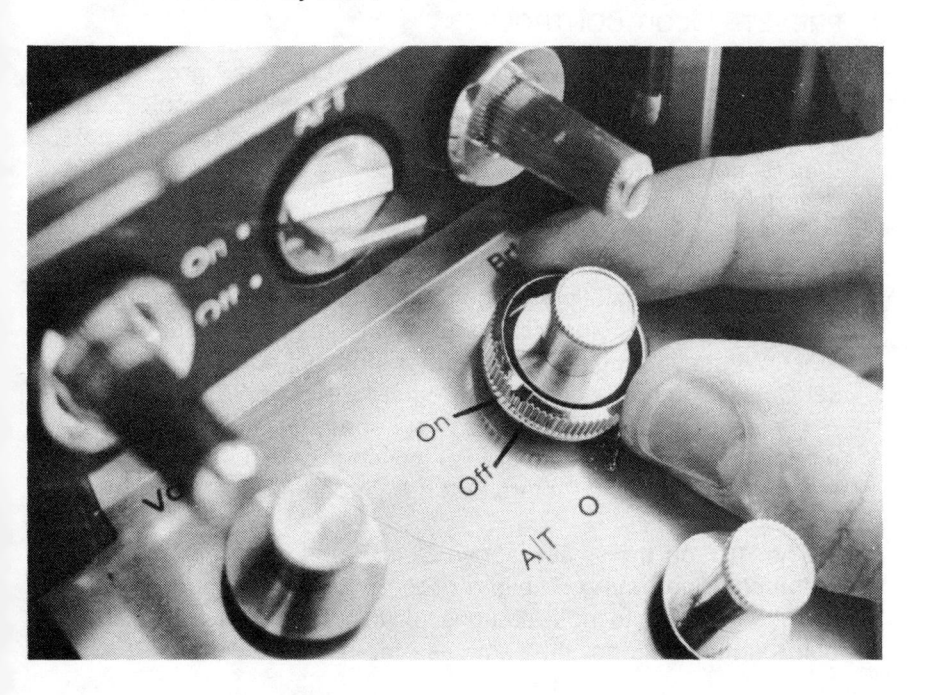

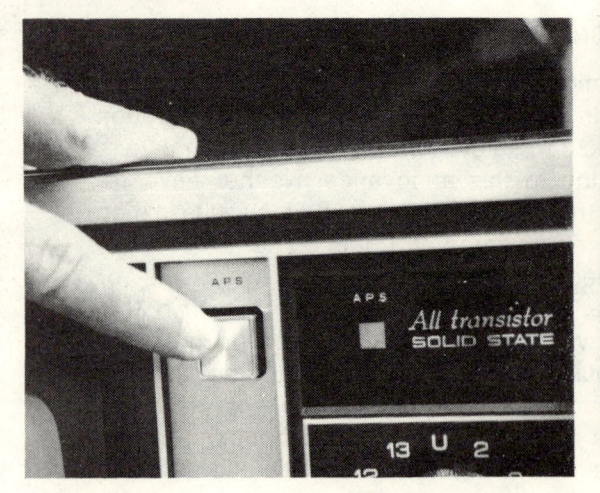

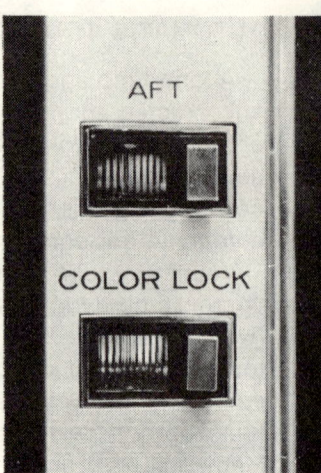

PRESET COLOR CONTROLS

One gimmick to overcome some of the unwanted green- and blue-face syndrome appears in several imported color receivers. Sometimes, they're even advertised as automatic; but they're not, in the corrective sense. They are more honestly described as *preset color controls*.

An extra set of Color and Tint controls are equipped with a switch (sometimes Brightness and Contrast are included). The technician who installs your color set adjusts these spare controls for the best color picture on whichever station you watch most. Whenever you change to other stations, you turn the Preset Color switch off and use the regular Color and Tint controls.

The photos show two such switches on imported color models. One bears the label APS, which the manufacturer says stands for Automatic Picture Setting. The other is labeled Color Lock.

The AFT on the switch above Color Lock is an acronym for *automatic fine tuning.* That's a properly automatic circuit most color sets need to hold stations exactly tuned for picking up color signals. It's not the same as atc.

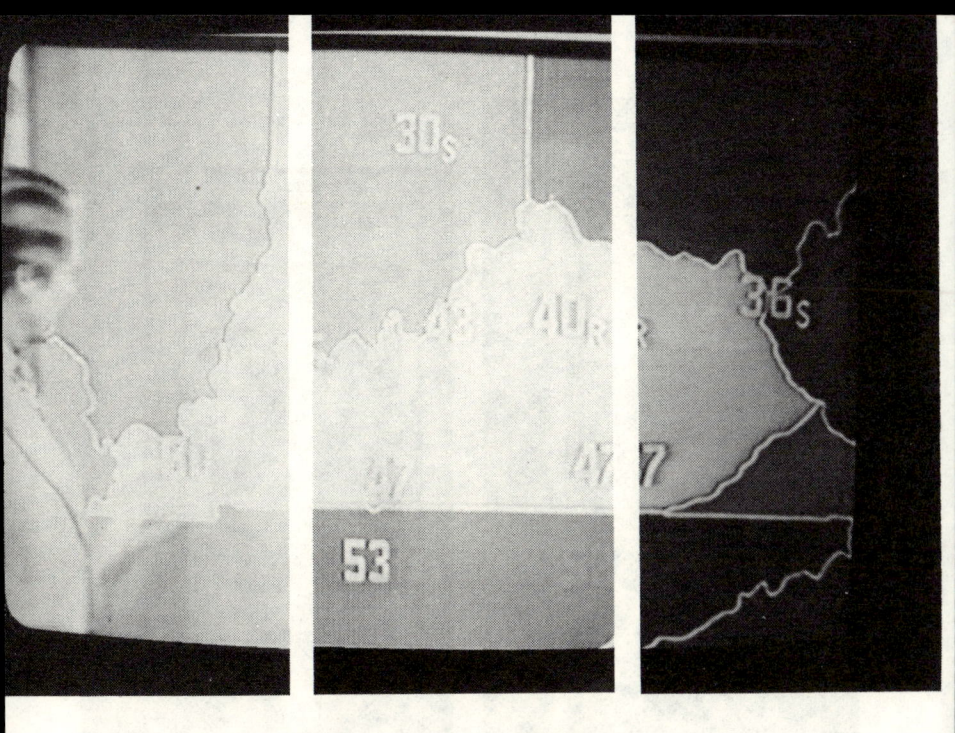

GRAY SCALE, CONTRAST, BRIGHTNESS

These words are intertwined to some degree with *video* and *luminance* (page 39). Begin with *brightness*. This refers to the illumination of a raster on the television screen. You can see results of turning up the brightness control whether there is a picture or not. This knob works the same on color sets and monochrome.

Contrast refers to the ratio between blacks and whites. Technically, contrast depends on how strong you make the video being fed to the picture tube. The more video, the blacker the blacks and whiter the whites. The Contrast knob is the one that controls this, the same as for a black-and-white set.

Gray scale is a term used only with color receivers. It describes whites, grays, and blacks in a monochrome picture viewed on the color picture tube. (You can make any color picture black-and-white by turning down the Color knob.) With Brightness and Contrast knobs set correctly, you should see a full range of tones from black to white, with plenty of light and dark gray levels in between. This gray scale is important to color rendition during color shows. There should be no slight shades of color in the picture as you turn brightness from overbright to too dark—as in the photo. Later, you'll see much more about gray scale.

43

UHF, VHF

You probably know these initials from black-and-white television. They stand for *ultrahigh frequency* and *very high frequency*.

Ultrahigh frequencies are the operating signal frequencies of stations with numbers above 13. There are 70 uhf channels, from 14 through 83. Almost all television receivers now receive both uhf and vhf stations; all sets sold since the early 1960s were required to have uhf tuners. Pretty soon, uhf and vhf tuners will be alike. The Federal Communications Commission, which oversees such things, has decreed this. Indeed, some manufacturers have already introduced pushbutton tuners (opposite page) in which you can't tell uhf buttons from vhf except by labels. Others have various contrivances to make uhf tuning easier.

Very high frequencies cover twelve channels of television: 2 through 13. Some industry forecasters predict that sometime these older frequencies will be eliminated for television and everything will go to uhf. That's no time soon, so don't fret about it. Just be sure any color receiver you buy can pick up both uhf and vhf. Preferably, uhf should tune as easily as vhf.

VARACTOR TUNER, DIODE TUNER

The ruling by the Federal Communications Commission that all tv sets must eventually tune as easily to uhf as to vhf has forced some innovations by manufacturers. One of the newest is the varactor tuner.

A *varactor* is a tiny electronic diode that can be made to change the frequency in a tv tuner just by a voltage variation. A television tuner like this, often called a *diode tuner,* goes from one channel to another without the big clumsy switches and drums you're accustomed to. The Sylvania version in the photo above has eleven small pushbuttons. You just punch whatever channel you want. No bump-bump-bump; the station appears instantly.

The varactor tuner in this form has drawbacks. You can tune the color receiver only to whatever stations are "programed" into the tuner. If you move, or another station comes on the air, it takes a trained technician to adjust the push buttons for the new channel or channels. But varactor tuners in various forms are destined to grow more popular.

PINCUSHION

This is one of those terms that doesn't sound as if it belongs to color television. It won't creep into ordinary conversation. But it's an odd one you can remember for later, in case you ever have to describe the condition to a service technician.

Pincushion, or *pincushioning* as it's more familiarly called, relates to the shape of the color-tv raster near the top of the screen. In the photo, the picture is down some, as if centering is off. But it's that bowed shape that is given the name "pincushioning."

Special correction circuits in most color chassis allow your technician to adjust out any abnormal pincushioning. Keep the appearance in mind. Lines across the top could be bowed, even if not bad enough to show the television picture framing (as this photo does).

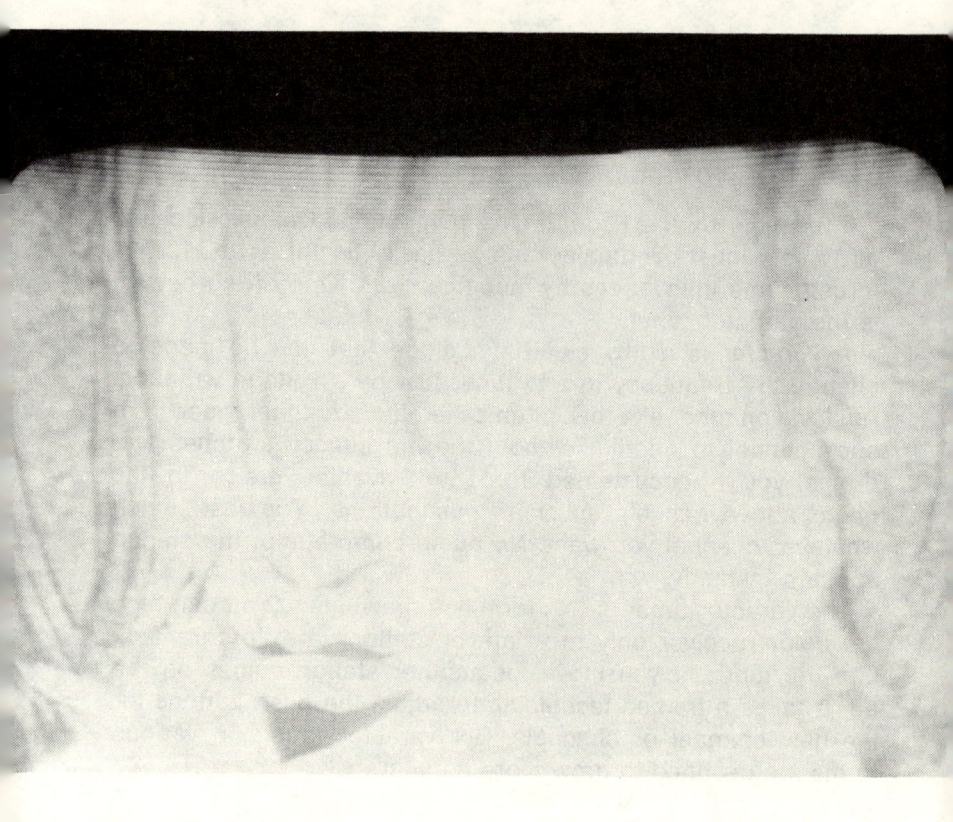

Chapter 4

Buying Your Next (or First) Color TV

Enough of that technical stuff. What you probably want to know most is how to go about finding your "dream" color set, the one that just suits you. There are so many different places to buy color, and so many different sorts of people to buy color from, choosing may be hard.

And what about afterward? Do you just lug the thing home on your shoulder? Not likely, unless it's a portable or a very small table model and you bought it without delivery or installation. That's one way to save money, but you have to decide whether the saving is worth the extra work and the chances you take.

The next several pages spell out some of your concerns in shopping for color television. What can you expect and what can you demand as your rights as a purchaser? Equally important are the limitations of buying at one place or another—and what things you have no right to expect.

Your object is to enter any store as well informed about color television as you can be. To that end, don't rush out after reading only this chapter. Go through the others, too. Know ahead of time whether you need a new antenna. Plan for the delivery and installation. Be aware what adjustments you can and can't handle yourself. Decide now what to do about service and repairs if and when you need them. In short, size up the whole picture before you plunge.

WHAT KIND OF TELEVISION STORE should you go to for your color set? You can find all flavors and sizes.

A high-volume tv store like the one above may have slightly lower prices. But the manager seldom backs his merchandise with any warranty other than what the manufacturer arranges for.

The neighborhood store owner (below) may offer prices no lower than any other store, but a good one generally stands behind the manufacturer's warranty with his own guarantee of service. That's often an important distinction. It's worth many dollars and saves lots of headaches if your color set happens to develop trouble. Chances are, you'll also get qualified delivery and installation.

Big discount stores, like the one at the right, often house large appliance departments. You can expect, and may get, price savings. You might get free delivery. Don't confuse that with free installation; it's not the same thing. A delivery man might remove the boxing covering and even plug the set in for you. But installation and proper adjustments come under the head of technical work and require a trained technician.

So, where you go depends on your own preference. The neighborhood store usually has the more friendly atmosphere, but not always. A few fast-selling tv stores offer service and warranty, but most don't. The discounter may have some very

low prices, but they might be for last year's model; the salesman should tell you. For that matter, any store may legitimately offer you a bargain on a last-year model. Consider it only if (a) the price is at least 20 percent below the price it sold for last year, (b) the set is new and still crated, and (c) the warranty stays the same as for this year's models. None of this applies to floor demonstrators; best avoid those.

The best bet is to shop them all. Pick out the kind of set you want, with the features you want, and then shop for price. Find out what extra is offered at the store that asks the higher price. That might be the real bargain, in the long run.

FINDING A COLOR TELEVISION STORE can begin as simply as opening your Yellow Pages to "Television Dealers— Retail." If you want to stay close to home, search out an address that's not far away. If you're looking for a particular brand, look over the display advertising for dealers who sell the kind you want. Don't be stampeded by display advertising, though. Some very minor companies take large phone-book ads. Be impressed only by what you see, not by what some advertisement says.

Phone a few stores. Get some idea how friendly they are. You can expect a sour reception in person if you get harsh treatment on the phone.

Don't expect anyone to quote you prices by phone. Few salesmen will do that. Besides, you should see what merchandise the quoted prices apply to. An attractive low price may not buy what you're really asking for.

Another way to find a set of the brand you want is to phone the distributor of that brand. His name usually appears under the brand name, where it says "For information call."

TAKE YOUR LIST WITH YOU SHOPPING. From your phoning, you should know fairly well where you want to try. Make a list and organize it so you can visit several stores in one area. That way you'll get to see more color tv sets in the same amount of shopping time. Leave room in your notebook, by the name of each store, for notes on what sets you see there, the prices, and what the prices cover.

Somewhere in your notebook, write down what features you've decided you want in a color set. List also the questions you want to ask every salesman: What guarantee? Who stands behind it? In writing? How much free service do you get if the set goes bad? For how long? Who pays for delivery? Who checks the color set at your house to make sure it's properly adjusted after delivery and installation? Who buys and puts up the antenna? For how much?

You probably have questions of your own to add, once you're out shopping. Add them, and then ask the same questions everywhere you go. Insist on specific answers, in dollars where feasible. Write them down. You'll pretty soon have a fair notion what you're going to get for your money wherever you buy.

VIEW MORE THAN ONE COLOR SET AT A TIME. Even in a small store, don't watch just one set. Look at the same picture on several color sets simultaneously. If there's a black-and-white handy, see how the same picture looks on it. That'll give you a good comparison that lets you spot abnormal shading in the overall picture.

One reason for viewing sets in bunches is this: The salesman might be able to cover up flaws in a set or explain them away, if that's the only set you're watching. It's tougher with four or five going. You can see what's station trouble and what's set trouble. The comparison makes a big difference in what you notice.

Furthermore, you can see if the store has installation and service people capable of keeping display sets up to par. If they can't keep the demonstrator sets in the store working, yours at home won't be any easier. Of course, if just one is on the blink, that's entirely plausible; demonstrators go bad the same as any others.

TAKE A LOOK AT THE CABINETS and have the salesman explain them to you. Even if you start out not knowing much about furniture, let the sales people you meet embellish your education. Ask about woods. Most nowadays are veneers. Find out why one kind of cabinet veneer is better than another.

Get the opinions of several salesmen. Ask a furniture expert before you go out color-tv shopping, if you know one. You may spend almost as much for the cabinet as for what's inside. It's a fact—the tv chassis you can buy in a simple table cabinet for $550 may cost you $850 or more in a fine-grained wood floor model.

Don't settle for a cabinet style that doesn't fit into your home decor. As much as with any other piece of furniture, you're going to be living with that color television for months—probably years—to come. Don't settle for one you don't like just to save a few dollars at a closeout sale. The cabinet's important. Spend as much time and care selecting it as you do getting the best insides.

IS IT EASY TO TUNE IN? Better make sure. Try the knobs out for yourself. Flip the vhf channel selector. Try out the uhf channels. (This may be meaningless in some stores; they use a master antenna system that converts all channels to vhf. Insist on trying out a station on the uhf tuner some way.)

Work the on-off button or knob. If it's the pull-on kind, see how easy the knob comes off in your hand. Twist the Brightness and Contrast knobs all the way up and down. Set them for a normal picture.

Finally, be sure to try out the color. If you don't know how, insist the salesman show you. Then *you* do it.

Survey what you've discovered. If the knobs are difficult to reach, you may get tired of getting down on hands and knees every time you want to change stations. If tuning is overly touchy, beware; don't get stuck with a color set you can't tune in easily. Don't be put off by a salesman who tells you that only the demonstrator is hard to tune in or adjust. Let him dig out a couple that tune easily, or you can assume all of that model are tough to tune.

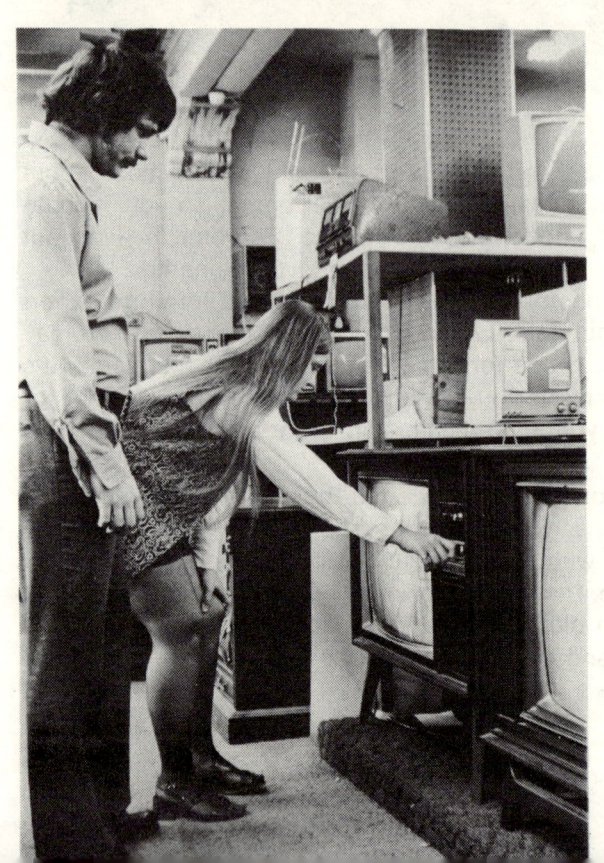

DO YOU PREFER REMOTE CONTROL? If you do, ask for a demonstration. Try the buttons out yourself. Discover how to work each function of the remote control. Is the control noisy? Is operation noisy in the receiver? How far away does it work? Find these things out for yourself before you buy.

Try several different brands and models of remote-controlled color sets. Some offer more functions than others. Some are quieter than others. Some cost more than others.

Make a list of the ways one kind of remote control works, and compare it with others. You probably won't choose a tv set entirely on the basis of its remote control, but that might be a deciding factor. Also ask about the different kinds of remotes that go with a particular brand. Some makes have more than one system, at varying prices. A less expensive one might serve your purpose just as well as an elaborate one.

WARRANTY ON A NEW COLOR TV is an important factor in your final choice. First of all, know one thing: The manufacturer of a new color television receiver guarantees all the parts inside the set for some period, from 90 days to a year.

Most guarantee the color picture tube for a year to 3 years. Longer guarantees may be prorated. That is, you might have to pay part of the price of a replacement.

Be sure you see *in writing* exactly what the manufacturer's guarantee covers. Do this *before* you sign any papers saying you will buy a color receiver or that you even intend to.

If a part goes bad during the manufacturer's warranty period, you may still have to pay for a service call, for pickup and delivery of the set to be repaired, and even for the actual repairs. That's nearly always true of long warranty periods. If so, you should know all this, in detail. The salesman should be glad to write it out for you or give you printed material that explains the warranty and its terms. A few color-set manufacturers *include* certain labor and repairs in their warranties, usually under special conditions. You should insist on knowing all those conditions *in writing* before you buy any color set.

Finally, some top-notch dealers back up the manufacturer's warranty with a service-labor warranty of their own. Find out what it is, what it covers, and under exactly what conditions you have to pay anything for warranty service. *Get it all in writing,* for a spoken warranty is the same as no warranty.

Chapter 5

Antennas Improve Color Reception

There are more different color-television antennas than there are color tv's. Selecting one that gives the best reception for the dollar spent is nigh impossible. There's so much of a technical nature bound up in finding a good tv antenna.

Some salesmen will tell you the antenna built into the set is enough. Others just don't want to talk about it. The best ones usually avoid committing themselves about antennas.

Take this one piece of advice: If you want the very best color tv reception, buy an antenna and have it put up outside (unless you get television by community antenna cable). The chart on the next page offers some suggestions on what kind.

RECOMMENDED TV ANTENNAS FOR VARIOUS SIGNAL AREAS

MANUFACTURER	VHF ONLY Local Signal	VHF ONLY Medium Signal	VHF ONLY Fringe Signal	UHF ONLY Local Signal	UHF ONLY Medium Signal	UHF ONLY Fringe Signal	VHF-UHF COMBINATIONS VHF LOCAL SIGNAL / UHF Local	VHF LOCAL / UHF Medium	VHF LOCAL / UHF Fringe	VHF MEDIUM SIGNAL / UHF Local	VHF MEDIUM / UHF Medium	VHF MEDIUM / UHF Fringe	VHF FRINGE SIGNAL / UHF Local	VHF FRINGE / UHF Medium	VHF FRINGE / UHF Fringe
Antenna Corp. of America	AC505 $9.95	AC511 $23.95	AC525 $62.95	AC320 $7.95	AC310 $14.95	AC315 $16.95	AC711 $10.88	AC712 $25.50	AC720 $39.95	AC710 $20.95	AC717 $29.95	AC725 $49.95	AC720 $39.95	AC725 $49.95	AC730 $64.95
Antennacraft	CS-500 $12.95	CS-700 $29.95	CS-1000 $69.95	Y-11G $9.95	Y-20G $14.95	Y-28G $19.95	BS-8 $9.95	CDX-650 $24.95	CDX-750 $39.95	CDX-750 $34.95	CDX-850 $44.95	CDX-1050 $69.95	CDX-1050 $69.95	CDX-950 $54.95	CDX-1050 $69.95
Blonder-Tongue Labs	0610 $23.95	0611 $31.50	0613 $46.95	3518 $8.45	0511 $15.70	0512 $27.20	0711 $31.50	0711 $31.50	0712 $38.95	0713 $46.50	0713 $46.50	0714 $55.50	0718 $66.95	0718 $66.95	0719 $78.50
Channel Master	3615 $18.75	3612 $50.75	1210 $84.95	4305 $10.95	4304 $16.50	0512 $25.65	3624 $15.95	3626 $22.75	0712 $32.95	3665 $35.50	1252 $53.10	1251 $60.30	3662 $75.75	3661 $89.50	1211 $99.95
Finney	CS-V5 $21.30	70-V17 $39.95	70-V28 $74.95	CS-U1 $12.05	CS-U2 $18.35	CS-U3 $26.40	CS-A1 $23.50	CS-A2 $28.50	CS-A3 $37.15	70-18A $56.95	70-18B $62.95	CS-C3 $72.35	70-23A $73.95	70-23B $80.95	70-29B $100.00
Gavin Electronics	1011 $31.50	1019 $49.95	1026 $65.95	CR-5 $7.75	CR-5 $7.75	CR-10 $11.95	1106 $20.95	1113 $32.95	1122 $55.50	1110 $25.50	1118 $41.50	1122 $55.50	1118 $41.50	1122 $55.50	1134 $76.90
GC Electronics	32-706 $16.52	32-709 $24.84	32-719 $41.49	32-8965 $3.75	32-8978 $7.91	32-8978 $7.91	32-1300 $8.78	32-507 $16.52	32-511 $24.64	32-1302 $26.30	32-1200 $27.10	32-519 $41.49	32-1202 $49.95	32-524 $59.94	32-1204 $74.95
Jerrold Electronics	VIP-301 $18.95	VIP-303 $37.25	VIP-306 $67.95	PAU-450 $11.75	PAU-700 $17.50	PAU-900 $28.95	VU-931 $25.75	VU-932 $34.25	PXB-48 $33.75	VU-932 $34.25	VU-933 $47.25	PXB-65 $53.75	VU-934 $58.75	VU-936 $73.25	VU-937 $125.00
JFD Electronics	LPV-TV40 $17.65	LPV-TV80 $37.60	LPV-TV190 $91.95	LPU-CTC-15 $17.50	LPU-CTC-21 $23.55	LPU-CTC39 $40.10	LPV-CTC110 $21.70	LPV-CTC220 $28.55	LPV-CTC220 $28.55	LPV-CTC323 $36.20	LPV-CTC426 $48.65	LPV-CTC532 $61.15	LPV-CTC639 $73.00	LPV-CTC639 $73.25	LPV-CTC747 $85.50
Kay-Townes	CP-5G $11.84	CP-15G $30.20	CP-36G $82.75	C-1G $11.15	UHF-4BT $8.03	PRO-51UG $50.25	CPG-5G $8.50	CPC-8G $13.47	C-11G $17.58	CPC-9G $15.25	CPC-12G $20.25	C-15G $23.98	C-21G $33.57	CT-34G $56.40	CT-42G $68.60
Lance Industries	LC-880 $20.30	LC-881 $29.90	LC-884 $59.80	KW4S $7.95	LU-820 $13.65	LU-840 $34.35	LC-30 $17.20	LC-80 $19.05	LC-81 $27.65	LC-37 $23.00	LC-82 $32.00	LC-83 $58.25	LC-38 $34.95	LC-39 $44.95	LC-119 $62.40
RCA Parts & Acces.	3BG09 $17.30	3BG13 $24.95	3BG27 $66.95	2BGD4 $4.85	7B140 $9.45	7B141 $12.45	4BG13 $21.95	4BG20 $31.95	4BG30 $53.25*	4BG23 $42.50	4BG23 $42.50	4BG30 $53.25	4BG36 $68.95*	4BG36 $68.95	4BG69 $125.00
RMS Electronics	STP-7 $15.95	STP-11 $21.95	STP-28 $59.95	COR-1 $8.95	SK-15 $10.85	SK-19 $12.95	BJ-8 $11.95	BJ-12 $19.95	SK-719 $50.95	BJ-11 $15.95	BJ-12 $19.95	BJ-15 $26.95	SK-1916 $71.95	SK-1919 $77.95	SK-1919 $77.95
Winegard	SC-500 $24.95	SC-520 $37.25	CW-2002 $100.00	U-965 $19.05	U-975 $26.33	U-995 $38.63	SC-790 $29.45	SC-800 $34.45	SC-820 $56.35	SC-800 $34.45	SC-810 $43.85	SC-820 $56.35	CW-960 $59.55	CW-980 $75.75	CW-1001 $100.00
Zenith Sales Co. Div.	973-83 $21.50	973-85 $44.50	973-87 $69.50	973-101 $9.95	973-8 $14.95	973-10 $31.95	973-89 $24.50	973-90 $32.95	973-91 $38.95	973-92 $54.50	973-92 $54.50	973-92 $54.50	973-93 $65.95	973-93 $65.95	973-94 $89.95

*Separate uhf, vhf antennas more economical.

Note: Prices are suggested list, not firm selling prices; they also vary with locality.

Courtesy Popular Electronics magazine

THE COLOR-TELEVISION ANTENNA CHART facing this page was compiled from information given by the manufacturers listed. You'll find antenna model numbers and prices. The antennas are grouped according to what kind of stations you have near you (uhf or vhf) and how far from the stations you live. Distances relate to signals about like this:

Location	UHF	VHF
Local	10–15 mi	10–20 mi
Medium	15–30 mi	20–50 mi
Fringe	30–60 mi	50–100 mi
Far Fringe	70 or more mi	100 or more mi

Just suppose you can buy Finco (Finney Co.) antennas nearby. You live 45 miles from the nearest stations. One channel is uhf and three are vhf. Consult the charts.

You need a uhf-vhf combination because you want to pick up both kinds of stations. Your mileage puts you in the medium-signal distance from the stations for vhf, but in the fringe for uhf. Look under "Vhf-Uhf Combinations" in the chart. Find the grouping for "Vhf Medium Signal." Under that, find the column for "Uhf Fringe Signal." Down that column, opposite Finney, is the antenna for you to order. It's a Model CS-C3, and its price is in the vicinity of $70. You can go to your antenna dealer or distributor and know exactly what to ask for.

If you are in rocky, hilly country, put yourself in the fringe category for both signals. You'd pick a stronger antenna. If you happened to want a JFD antenna, and needed to pick up fringe uhf and fringe vhf signals, you'd select the LPV-CTC747 model. Price is $85.50, but you're buying a stronger signal for your color receiver.

The chart offers a selection for long and short distances, so you don't spend more money than you have to. The pages that follow illustrate several antennas, show their styling, and list some characteristics.

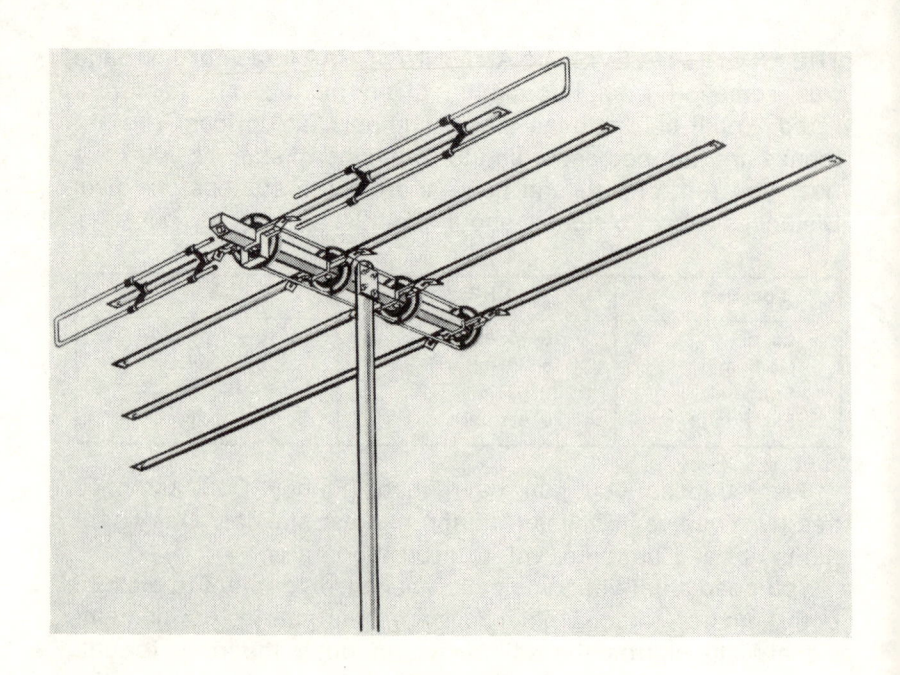

COLOR ANTENNA FOR SHORT DISTANCE

This one is made by Jerrold Corporation and goes by the name "Paralog-Plus." It has only one folded dipole element, the one in front. That's called the *active element,* since it connects directly to the downlead that runs to the television receiver.

The elements that look like straight rods are called *reflectors.* They reinforce the sensitivity of the front element, increasing what is called the "gain" of the antenna. This antenna has more gain—it picks up more signal—than a single dipole all by itself.

But the few elements mean that the gain of this model is low compared to that of big antennas. The small size and gain are plenty for stations that aren't far away.

The lengths of the elements in this antenna also stamp it as for vhf only—channels 2 through 13. As you'll see later, antennas meant to pick up uhf, too, have a string of shorter elements up front.

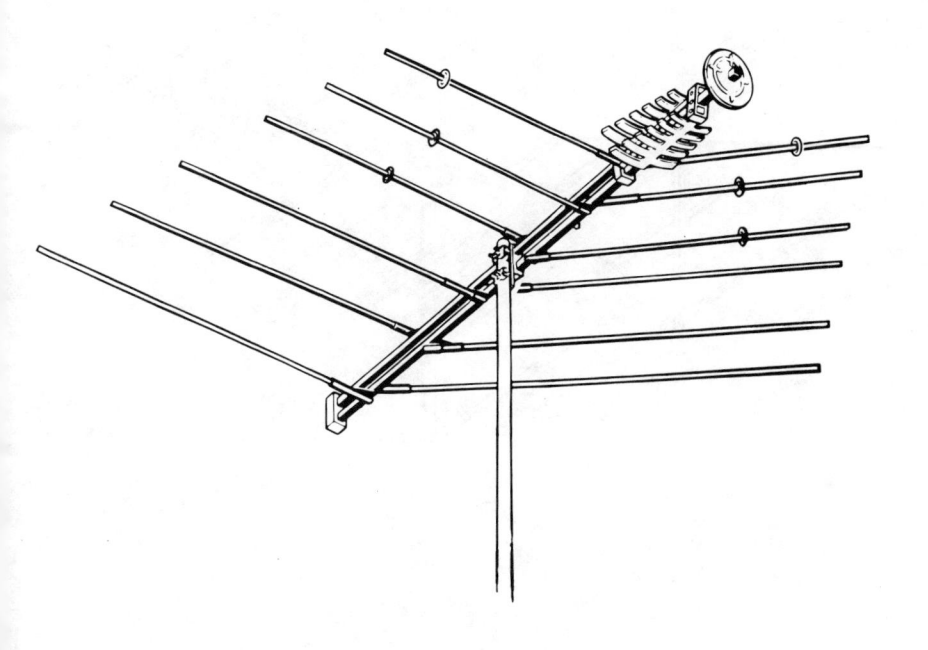

URBAN/SUBURBAN COLOR TV ANTENNA (UHF and VHF)

More elements make the most noticeable characteristic of this antenna compared to the one on the opposite page. That means it's stronger. The additional elements give it more "gain." It can pick up signals from farther away.

The second noticeable difference is the addition of some funny-looking stuff up front (the shorter dipoles or rods are always toward the front of the antenna). Those little flat-metal dipoles and that disc make up a special uhf antenna. It mounts right on the vhf antenna. The same downlead carries both uhf and vhf signals to the color receiver.

This antenna is made by JFD Electronics Corp. for use when you live a few miles from the stations, but not really in the fringe. The styling is called "Color Laser Log Periodic." The dipoles with the little circles on them are the active ones; the others (the back six rods, which form three reflector dipoles) merely reinforce the active elements. The reflectors also add directionality; they help the antenna shut out stations from any direction but in front. Sometimes that's an excellent quality for a color tv antenna.

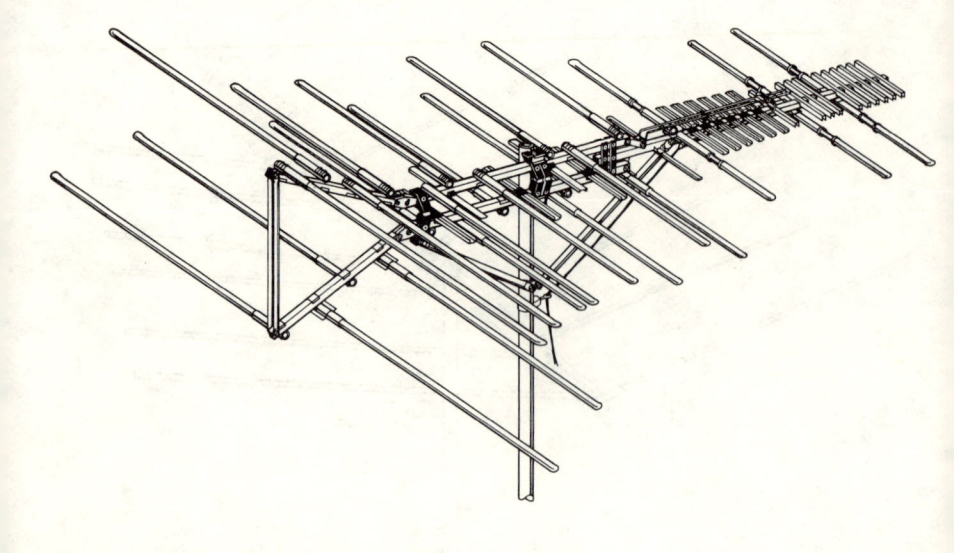

UHF-VHF COLOR ANTENNA FOR FRINGE AREAS

Notice how many elements this sensitive antenna has. Even over distances far away from the station, this Finco antenna picks up plenty of signal for a good color picture. The color receiver needs a better signal than a black-and-white set, and a powerful antenna like this one can deliver more.

Up front, an array of very short elements picks up uhf signals. That ability makes this antenna good for fringe uhf as well as fringe vhf signals.

A distinctive characteristic is the formation of the reflectors at the rear of this antenna. If you study them closely, you can see they're arranged on a delta-shaped boom at the back of the main twin-boom. This is a Finco innovation to improve directionality of the antenna and yet broaden its signal-capturing ability for color television reception. (That last is called *broadbanding*.) This Delta reflector is exclusive with Finco.

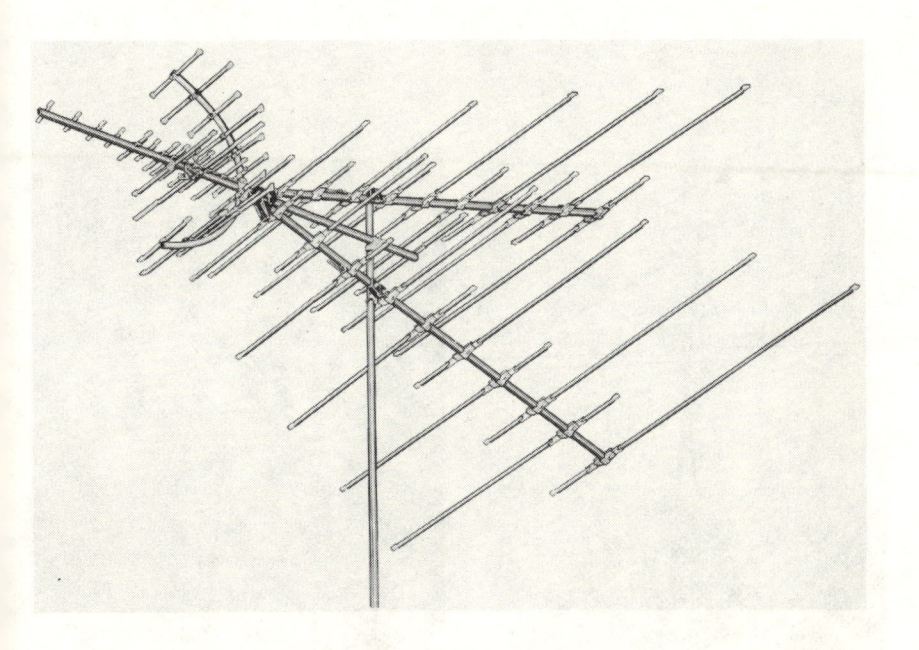

COLOR RECEPTION IN THE FAR FRINGE

Many, many dipoles, reflectors, and directors make this one of the most sensitive color tv antennas on the market today. This one is made by Winegard Co. and is called a "Super Colortron Color Wedge." (Of course, other companies make far-fringe antennas and Winegard less sensitive antennas for short-distance use.)

The short elements up front are directors and reflectors that build up sensitivity to uhf signals. This one can pick up viewable uhf signals 70 miles or more from the station, unless high hills intervene. The vhf elements pull in enough signal to watch from as far away as 125 miles. (Again, that depends somewhat on terrain.)

You don't need to spend this much money (price is $100) on a color tv antenna unless you're in a very bad reception location. But if that's what you need, it's fruitless to try to pick up good color with any less of an antenna. Whether you're 100 miles from the nearest tv station or living in the shadow of the transmitting tower, the picture you see on your color set is only as good as the signal fed into your receiver's antenna terminals.

FOR UHF RECEPTION ONLY

The top portions of the photos here illustrate what you can use for color reception if your only stations are uhf. Not many areas of the country have only uhf; most have a mixture of both kinds.

Or, you may live in a spot where vhf stations are in one direction and uhf in another. Sensitive tv antennas are directional. If you use a uhf/vhf combination, you'll need some way (an electric rotator will do) to turn it from one direction to the other. Or, you can use a vhf antenna pointed one way and a separate uhf antenna pointed the other as shown above.

One disadvantage of two antennas is that you need two lead-ins. Still, the important thing is getting a decent color signal to your color receiver. If it takes two antennas and two lead-ins, or whatever, better do it. A half-hearted color signal will never give you satisfactory color-tv viewing.

SHORT-RANGE INDOOR/OUTDOOR ANTENNAS

A new breed of electronically amplified antennas has been introduced to the color television market. These two have an ultramodern air about them. They can be used indoors or out and work fine when signal conditions are good. They are an improvement over the old set-top rabbit ears or the built-in monopole (a rod or two sticking up from the back cover).

The long wing-like antenna at the top is the "Sensar" from Winegard. The one with the flying-wing appearance is the "Stellar 2001" made by JFD. Both work for vhf and uhf, although experience shows (1) they don't pick up uhf more than 30 miles or so with any real clarity, and (2) they don't have much directionality and therefore can't eliminate ghosts. (More about ghosts later in the book.)

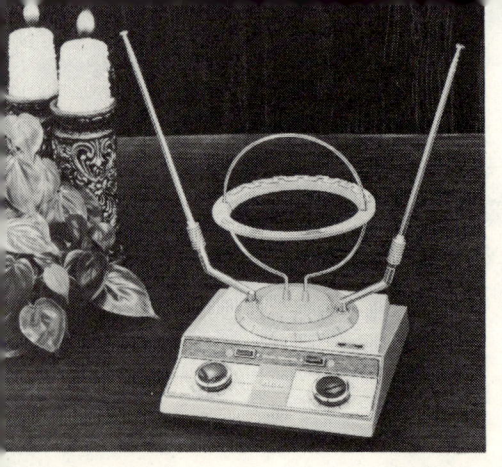

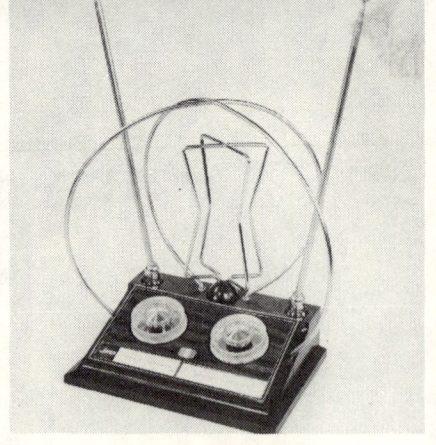

ANTENNAS FOR THE TOP OF THE RECEIVER

If a color-tv salesman tells you there's no need for an outdoor antenna, the chances are slim that he's right. Sure . . . you can get by without one, just as you can get by without watching a good color picture. But if you spend that much for a color set, you are entitled to good viewing. And the fact is, there are very few locations where you can watch a good color picture without an outside antenna.

For those exceptions, in localities where the television signal from stations is very strong and not bothered by multiple reflections (ghosts) from hills and buildings, antennas like those pictured on this page can do a suitable job. The only way you'll know is to try one. If the salesman insists that all you need is one of these, ask him to loan you one (*not* sell it and promise your money back if it doesn't work). The antenna that gives a good black-and-white signal may be inadequate for color. But if you can get by with an indoor antenna, here are some that do well. The one above comes from RCA; below left, from JFD; below right, from Jerrold. Beware of cheapness among set-top antennas; buy a brand whose name you know.

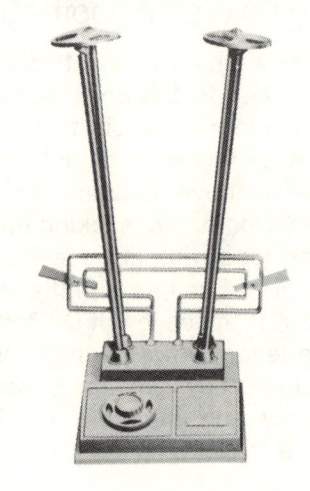

THE STORY OF COMMUNITY ANTENNA TELEVISION (CATV)

The importance of a strong signal was acknowledged even before color became so popular. In some out-of-the-way places, citizens put powerful antennas on top of tall, expensive towers and supplied the signal they picked up to other citizens around town—for a fee, of course. The signals were distributed, as master antenna signals are, by coaxial cable. Community antenna television (CATV) systems have proliferated and are known today by their distribution method: they're called *cable television*. More than 2500 such cable systems exist today, some serving thousands and thousands of people. Small systems serve only a few hundred.

More often than not, the cable brings you good signals—plenty for good color reception. But poor cable tv systems are plentiful enough that you might opt for your own outdoor antenna.

You can judge to some extent by the kind of picture you see on black-and-white. Look for multiple images. If you see them a little on monochrome, they'll be much worse on color. If there's snow in black-and-white pictures, your color reception will be marginal at best. Maybe you're in a location where even poor cable signals are better than you can do with an antenna outside. If so, you may be disappointed with television, unless the cable company does a really good job.

WEATHER SENSOR ⇩

⇧ CATV TOWER

⇧ CATV HEADEND "SHACK"

COMMUNITY ANTENNAS ON WATER TOWER ⇨

68

LINE AMPLIFIERS AND DROPS

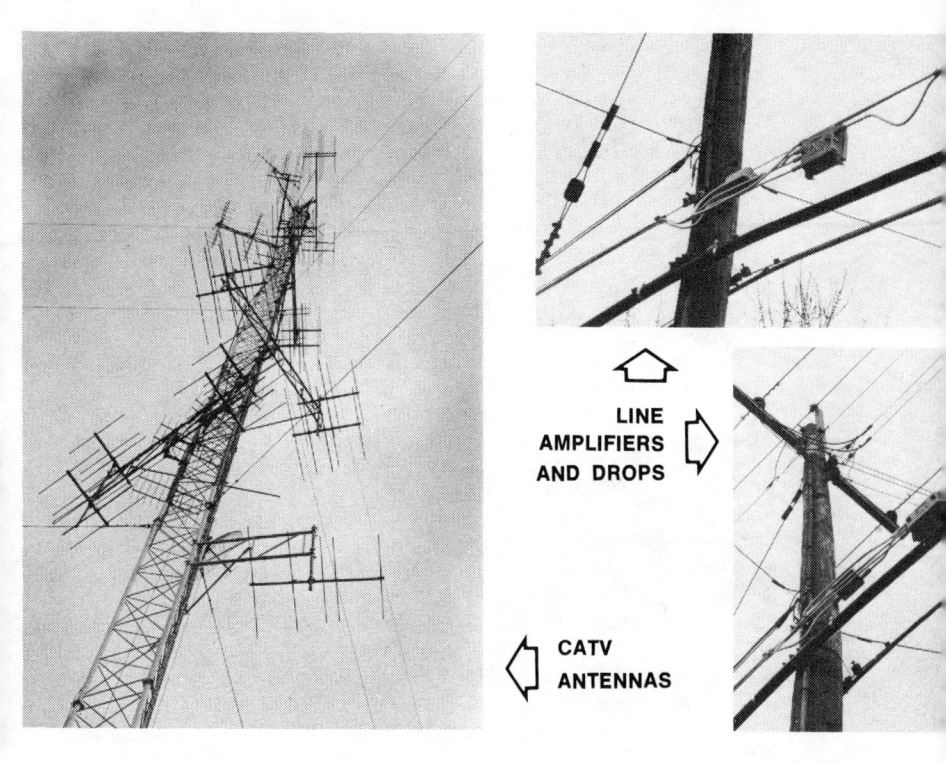

CATV ANTENNAS

ONE SECRET TO SUCCESSFUL CABLE TV is a tower filled with powerful antennas. The tower should be on the highest ground around or on some tall structure. Signals from all those sensitive antennas go down to a rack of amplifiers in a small building at the base of the tower. Uhf signals are converted to vhf. Then all the channels are distributed by cable to tv-owners who pay the monthly fee.

All twelve vhf channels can be used. With special equipment, some systems aim for 20 or more channels for every subscriber. Many cable operators enhance their system by installing extra services on unused channels. A bank of weather gauges and a small monochrome camera display time, temperature, humidity, wind velocity and direction, and special announcements all day long. A few, often using local high school or college students, originate their own programs on one channel—although seldom in color.

Industry seers predict the day when all television will come to you by cable instead of over the air. Cable systems will interconnect about like telephone networks do today. But that's not right away.

IF YOU LIVE IN A MODERN APARTMENT BUILDING, you probably get your television signal from some kind of master antenna on top. It feeds its signals to special amplifiers, and a network of cables carries the signals to all the apartments. You have to take whatever signal is delivered to you. On the roof, a broadband antenna like this Channel Master "Super Vector" could be the pickup device.

The antenna shown here is without uhf, for an area in which there are no uhf stations. Many apartment-type antenna systems pick up whatever uhf stations are available and convert them to unused vhf channels. That's because vhf is easier to carry by cable than uhf is.

If round (coaxial) cable is the distribution means instead of regular flat lead-in, you need a special transformer to couple the signal from the coaxial cable to the two-screw antenna terminal on the back of the color set. Manufacturers of some modern receivers, knowing that many receivers today connect to coaxial cables, include that kind of connector (inset) with their sets.

Chapter 6

Good Installation Avoids Problems

Old or new, your color television receiver ought to give you the very best color picture of which it is capable. And you should have a very minimum of difficulty and breakdowns. That's an ideal. One sure approach to that ideal is through careful and proper installation.

Too often, delivery men just plop down a color set on the living room carpet, plug it in, and take off. Some don't know any more to do and shouldn't be expected to. They're not trained for that.

But that's not the way a new color tv should be treated. It deserves some attention. Just as a new car needs a tuneup before you drive it out, a color set right out of the box works better after certain adjustments are made. Sometimes, a repeat checkout and adjustment (called a *setup,* in the trade) makes a world of difference after the set is a couple of months old.

And there's that matter of a color tv antenna. Just jamming a powerful antenna on a pipe and clamping it to the chimney isn't enough. The color set works best only when fed an adequate television signal. Installation should include making sure you have a good signal on all stations. That's the bailiwick of a competent technician, not the truck driver who delivers.

Here are some of the things you can watch for to help yourself be assured of satisfactory color tv viewing.

DELIVERY BEGINS EVEN BEFORE THE BOX is uncrated. At the really concerned television store, the crate was unsealed the day before delivery. The box was slipped up and off; the set was plugged in and turned on, and allowed to run a few hours to "cook out." That way, anything not correct is likely to show up before you try to turn the set on at home. If a trained technician is part of the staff, he made any adjustments the set needed. It's a rare set nowadays that works perfectly right out of the box.

Then the box is turned back down over the set for delivery. When it gets to your driveway, the set that's been checked out for you probably has the box flaps unstapled. The delivery men open up the bottom of the carton and take out the bolts that hold the wooden packing frame. With the frame out of the way, they turn the set over—right side up, now—and lift off the shipping carton. They'll take that big box with them unless you want it for the kids to play in.

WHEN IT'S CARRIED IN THE DOOR and set down in its corner is the time to go over the cabinet carefully for scratches and mars to the finish. Lodge your complaint for any damage immediately. Don't wait till evening or the next day. Damage noted while the delivery men are still there gets more attention from the store; point out any flaws to them. If there is serious damage, don't accept the set. Have it returned to the store. Once you accept it, you may well be stuck with it.

But help the delivery men all you can. Hold the door. Keep the dog and cat and the youngsters somewhere out of the way. It's unfair to hold delivery men responsible for damage you make impossible to avoid.

INSIST THAT SOMEONE SHOW YOU AGAIN how to operate that particular model of color television. When you picked it out at the store, the salesman should have explained as much operation as you needed. But delivery day is later. You need a refresher.

Maybe the delivery men are familiar enough with the set to instruct you. They should be. Better, of course, is when the set is delivered by a team made up with at least one trained technician.

Even then, this customer instruction step is often omitted. Yet it is as important as any portion of the delivery procedure. Insist on it. You're entitled to it. Your new color receiver will operate better, you'll need fewer service calls later, and you can relax and enjoy viewing if you know the knobs are set right.

YOU CAN DO YOUR OWN DELIVERY AND INSTALLATION if you know how. You may save some money that way, but only if you do the job right. If a technician has to come out and clean up a mess you make, you won't have saved a red cent.

You've already seen a number of color television antennas. That's the place to start in putting together your own top-grade installation. You can install most antennas yourself if you don't mind climbing around on the roof. The box the antenna comes in carries a sheet of instructions showing how to put it together and attach the lead-in.

Good common sense tells you to mount the array on a sturdy support. You can clamp to the chimney if it's solid. Or, run a pipe from the ground to 10 or 20 feet above the roof. If you choose a simple pipe mast on the roof, use guy wires to keep it up straight.

Make sure any pipe can't twist, either. That may take clamps. Your local tv shop has them or can tell you where to buy them. If there's an electronic parts distributor nearby, the salespeople there can show you what materials you need and how to use them. Patch any holes you make in the roof for fastenings. Roof-calking compound works fine.

BE SURE TO GROUND the mast, so it can't collect lightning charges. Safety regulations stipulate an 8-foot copper rod driven into the ground. Pour water around it if you have trouble driving it in. But start at least a foot away from your house foundations, and be sure no sewer pipes cross the spot where you plan to sink the grounding rod. Use heavy copper or aluminum wire (not leftover guy wire) solidly bolted to the mast and run as directly as possible to the ground rod—no crooks or turns.

The inset photo on the opposite page shows one terribly poor ground installation. Mistake 1: It is connected to a gas standpipe. Not only bad but dangerous. Equally dangerous is a rooftop connection to a protruding soil pipe. Mistake 2: Tv lead-in twinlead is used for the grounding wire. It's much too small to be of any value for lightning discharges. Mistake 3: The clamp wraps around a rusty place on the pipe. A technician's test of this ground showed it to be utterly useless.

If you're using the older flat or oval twinlead, you may want to install a lightning arrester on the lead-in. The arrester usually comes with instructions. But be sure you buy the right kind for the lead-in you're using. The different types are not interchangeable. Take a 6-inch sample of your lead-in along with you when you go to buy the arrester. If your arrester installation is old like the one at the bottom of the page opposite, throw it out, even if you don't install a new one.

SURE BET: You can save a lot of work and exasperation by hiring a tv antenna specialist to do the installation for you. But insist that he install proper lightning protection.

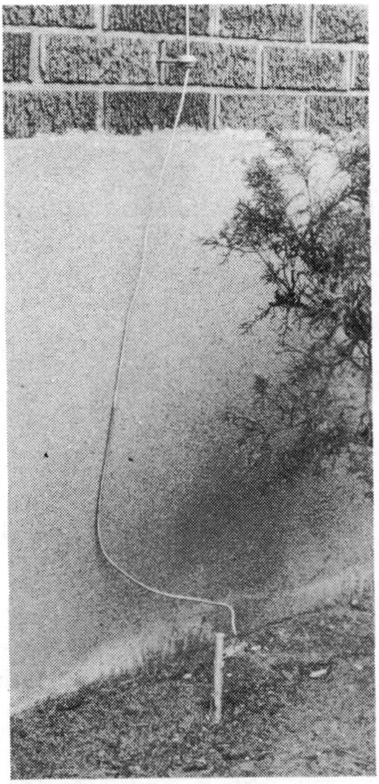

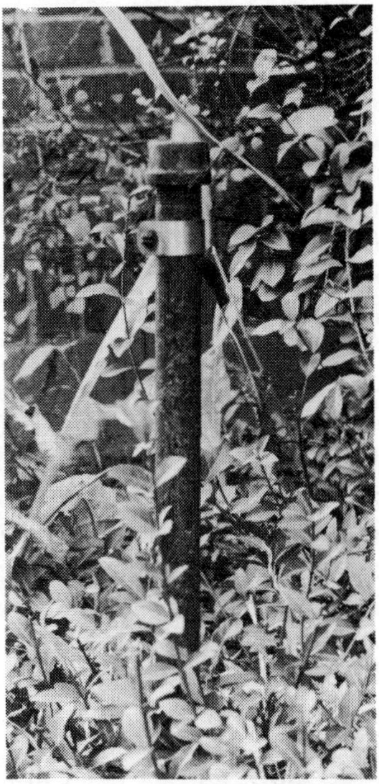

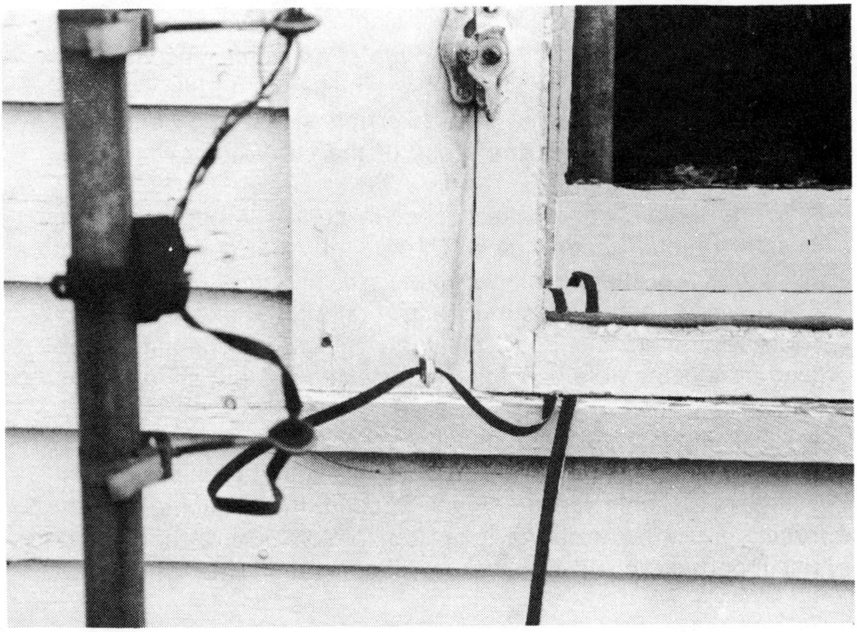

RUNNING A COLOR TELEVISION ANTENNA LEAD-IN takes more care than for black-and-white. That's not to say a *good* black-and-white installation won't work for color. A truly good one will. The point is just that you need a cleaner television signal for color than for black-and-white.

In older antenna installations, and in many poorly installed new ones, the lead-in wastes as much of the television signal as it delivers to the receiver. Sometimes, just putting in a new lead-in and running it properly from antenna to tv set makes a startling improvement in color reception.

For example, television lead-in should be kept away from the pipe mast, off the wet roof, clear of metal gutters, and not allowed to touch the wet sides of the house. The device for all that is called a *standoff*. You can buy any one of several varieties, but make sure it can hold your lead-in securely without pinching and without allowing metal to touch the surface of the lead-in wire.

Strap-type standoffs fasten to the mast and to downspouts. Wood-screw types lead the wire down the sides of the house and across the basement inside. Old though they are, the standoffs at the bottom of the preceding page are acceptable examples. (The one by the window should always be *below* the point of entry, so rain can't drain inward. Anyway, running lead-in under a window sash like that is taboo. Don't do it. Drill a hole beneath the window frame, slanted upward.)

The photos on the opposite page show some absolute no-no's. The top picture shows twinlead nailed along the weatherboarding of the house. Very bad. Terrible signal loss from nails; worse when wet. From the looks of the entry hole, water can drain in there, too.

In the middle photo, the tv lead-in comes through the wall beside a metal heating pipe. That's bad. Nearby metal drains away much of the signal, particularly if the lead-in carries uhf. The lead-in should be supported by standoffs and held away from any metal, including the metal electrical conduit. If the lead-in needs a hole through the basement wall, it should have its own.

The bottom photo is a closeup view of a very common source of antenna trouble. First of all, running the lead through a metal window is terrible for the signal. Second, the clamping usually breaks the wire or shorts it out completely. Third, the wire is unsupported. This lead-in was installed by the owner of a brand new color tv set. He put up with a poor picture for a year, all the time blaming the set. An antenna expert ran a new lead and put a beautiful picture on the tv set in one afternoon.

Don't you be responsible for a sloppy tv installation. Hire it done right, if you don't want to spend the time and effort yourself.

HERE'S WHAT THE ANTENNA-LEAD ENTRY should look like. You bring the lead down to a standoff mounted below the point where you drilled the entry hole. Clamp the lead-in so it still runs downward. Then curve it upward to the entry. The downward curve forms a *drip loop,* so water drops off instead of following the wire into the house.

The entry hole should slant upward. Any moisture that tries to collect drains back out down the lead-in wire. You can buy special long insulator tubes that guide the lead-in into the house. They're not necessary, though, unless the house has a metal shell. Even then, you can get by all right if you leave the drip loop as shown and don't run the lead-in along parallel with the surface of the metal siding.

Support the lead-in near the entry inside the house. A woodscrew standoff in a basement-ceiling (floor) joist does fine. Then route the lead-in so it avoids metal pipes. Don't make sharp right-angle bends in the lead-in wire either inside the house or out; that, too, cuts down the signal that reaches the color tv set.

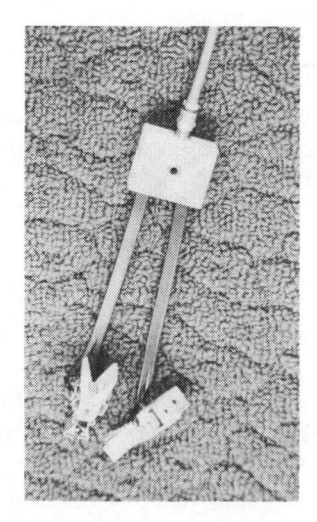

BEHIND THE COLOR TELEVISION RECEIVER, you may have to use a device called a uhf/vhf splitter. If your antenna picks up both uhf and vhf television signals, one lead-in carries both signals to the color set. Most tv sets at present have two pairs of terminals on the back, as the photo illustrates—one pair for the uhf tuner, one pair for vhf.

The uhf/vhf splitter separates the uhf and vhf signals right at the back of the tv set. It's the white device in the photo. You connect the antenna to the Input terminals of the splitter, and then fasten the uhf wires to the tv set's uhf terminals and the vhf wires to the set's vhf terminals.

You can manage without a splitter. A clothespin-type antenna clip on the end of the antenna lead-in facilitates moving the lead-in connection to the terminal for whichever channels you're watching.

The inset shows another kind of uhf/vhf splitter. This one comes from a recent type of color-tv antenna system that uses coaxial cable for the lead-in. That's the round wire you see. The coaxial wire has a plug that screws into the splitter. The output leads are equipped in this case with clothespin clips for easy connecting. They could just as well be wired to the screw terminals.

CAN YOU RUN TWO COLOR SETS from a single antenna? The answer is: Yes, if the antenna develops good strong uhf and vhf signals.

What you use is a *two-set coupler.* The antenna lead-in runs to somewhere about halfway between the locations of the two color tv sets. Then you connect it to the Antenna screws on the back of the coupler. Connect a lead-in to the Set-1 screws and route it to the first tv set, using the same precautions you follow with all the other lead-in. Connect another lead-in to the Set-2 screws, and route that lead-in the same way to the other color set.

The two-set couplers shown here plug into house power and amplify the signals sent on to the tv receiver. That way, there's no signal loss from dividing up the signal. Couplers are available that distribute amplified uhf and vhf signals to four or more receivers. Just be sure the coupler you buy handles both uhf and vhf; some don't.

Some two-set couplers are not amplified. They're all right, but each color tv gets less than half the total signal picked up by the antenna. That might not be enough for a top-notch picture. A color-tv technician or a tv antenna expert can measure with an instrument called a *field-strength meter* and tell you if you have enough signal to spread to two color sets. You need at least a thousand microvolts of signal for each set.

ADJUSTMENTS THAT SHOULD FOLLOW DELIVERY of a color television may be your responsibility. It all depends on what kind of deal you made with the salesman you bought from.

In the tv business, minor adjustments are called "a setup." When you buy a color tv right out of the box, no one has had a chance to check operation or set up the adjustments. More often than not, your brand-new instrument needs those adjustments. If they're not included in delivery and installation, you'll probably have to pay a technician to do them. They are NOT covered by manufacturers' warranties. Unless the dealer you bought from has agreed to set up the adjustments, you'd better just hope they don't need correction too badly.

One step included in the setup of some color brands is a process called *degaussing*. It means demagnetization. Occasionally in transit, or even in your home, the metal parts of the color picture tube pick up some magnetism. The result is blotchy-looking color on the raster when there's no station tuned in. If it's really bad, it fouls up color programs, too.

Degaussing is nothing you can do yourself. A technician does it with a special gadget called a *degaussing coil*. He plugs the coil into a power outlet, waves it magically in front of the set and around the sides, then moves it away and shuts it off. The set is demagnetized.

SOMETIMES THAT BLOTCHY COLOR in the raster comes from a different fault. The color picture tube has, as you remember, a deflection yoke and convergence yoke wrapped around its neck near the back. Sometimes these slip when a set bounces around. If they do, mysterious colors show up when there's no color picture. The colors look like smears of pink or blue or green, usually in one corner or along one edge.

Included with the deflection yoke is a small set of magnetic rings called *purity rings*. Their positioning is important to how purely white the raster stays. If they slip or turn, even slightly, your new color set may show those splashes of color in the white raster.

Your installing technician usually starts by degaussing the set. Then he makes these purity adjustments, aligning the yokes and the purity rings. If the raster still isn't pure white, he may have to adjust the Screen and Drive controls (remember from pages 15 and 16?). That's called "adjusting the gray scale." Adjusting yoke and purity rings is called "setting up purity."

CONVERGENCE MAY BE NEEDED in your new color set. If so, you'll know it by the very sharp red or blue fringing you see around the edges of white or black objects. The need for convergence is easiest to detect with a monochrome picture. A technician who wants to hide it may switch to a color program. Just turn the Color control off. If there's bad color fringing up toward corners or along the edges, a convergence procedure should be part of the new-tv setup.

This is a time-consuming process and will cost you if it's not part of your installation. The technician has to use a special signal generator that makes dots and/or lines on the picture tube screen in place of a picture. Then he makes some dozen or more adjustments, each one of which has to be exactly right. He has to concentrate, so keep the kids and animals away. You might consider not bothering him yourself.

Once set properly, convergence in modern sets stays pretty well. If you move the set around much, the job might have to be done again. Otherwise, you might want convergence—and all the other setup adjustments, for that matter—checked over every year or two.

THAT ALL-IMPORTANT AND VALUABLE WARRANTY should have been explained to you when you bought your color set—or even before, if you shopped wisely. Any warranty or guarantee that the dealer has promised must be in writing on the sales slip. Any guarantee not in writing is not valid. Don't forget that. No writing—no guarantee.

The manufacturer's warranty is something else. It may be loaded with limitations and gimmicks, but you can usually depend on getting whatever such a warranty promises. If the dealer doesn't get it for you, a letter to the manufacturer gets action.

There's one important catch. If the new-set warranty card isn't properly filled out and mailed within a specified length of time, your warranty there is no good either. The card comes in the box with the set. If it isn't there, don't accept the set. Some dealers fill out the warranty card and mail it for you. Better that you do it yourself; then you know it's done. Ask for help if you need it.

But no matter what, *get that warranty card in the mail.* It might save you a hundred or more dollars later.

Chapter 7

Tuning Your Very Best Color Picture

Suppose you've shopped diligently and finally bought just the right color television set for you. Not too big, not too small. Not too costly, not too cheap. You hired the best antenna installer in town to put you up a new super-saber color-grabber, and had him run the lead-ins with meticulous care. You waited to unpack your new receiver until a tip-top technician was right there to help pry out the packing staples. He spent one whole afternoon fiddling every adjustment right down to the last gnat's tooth. The picture looked beautiful.

And now the experts have all gone. It's more than a week later. There's just you, the family, and that beautiful new instrument. But something happened. One of the kids opened up the soapdish and tinkered with some of those knobs. And now you have a problem. How to get them back to normal? How to recapture the glowing color you learned to enjoy that first week?

Well, first of all . . . relax. Tuning in a color tv is much simpler than you might think. Sure, some of the knob-turning is more critical. And yes, there are a few more knobs to turn. And agreed, you even have to be more careful tuning the regular knobs—the same ones you had on the old black-and-white set.

But you can still relax because the whole process is not all that big a deal. There is a 1-2-3-4-5 sequence you can follow to perfect color. The worst difficulty comes when the inexpert color tv watcher (perhaps you, before now) tries adjusting all those knobs in haphazard order. Things seem to get worse instead of better.

This entire chapter is devoted to the five steps that put your color set right back like it belongs, *every time*. If these five steps don't work, you can bet there's something wrong with the set. Make sure you learn them in their proper order. Don't skip any step. They're so quick, and sometimes so simple, you might be tempted to jump around. Don't. Here are the steps so you can use this page as a reminder whenever you're tempted to vary the procedure. First turn the color down. Then proceed as follows:

1. Fine-tune the station precisely.
2. Steady the picture with horizontal and vertical holds.
3. Adjust Brightness and Contrast controls for the smoothest monochrome picture.
4. Turn Color up just enough to see color plainly.
5. Turn Tint knob for most pleasing flesh hue.

That's all there is to it. Sounds simple doesn't it? Really, it is. But each of the five steps has its own little tricks and idiosyncracies. Go over the pages of this chapter carefully. Then try the steps out exactly as shown, using your own (or a borrowed) color set. In no time flat, you'll be the best color-tuner-inner on your block. From there on, it's just sit back and enjoy the best color your set (or the neighbor's) can give.

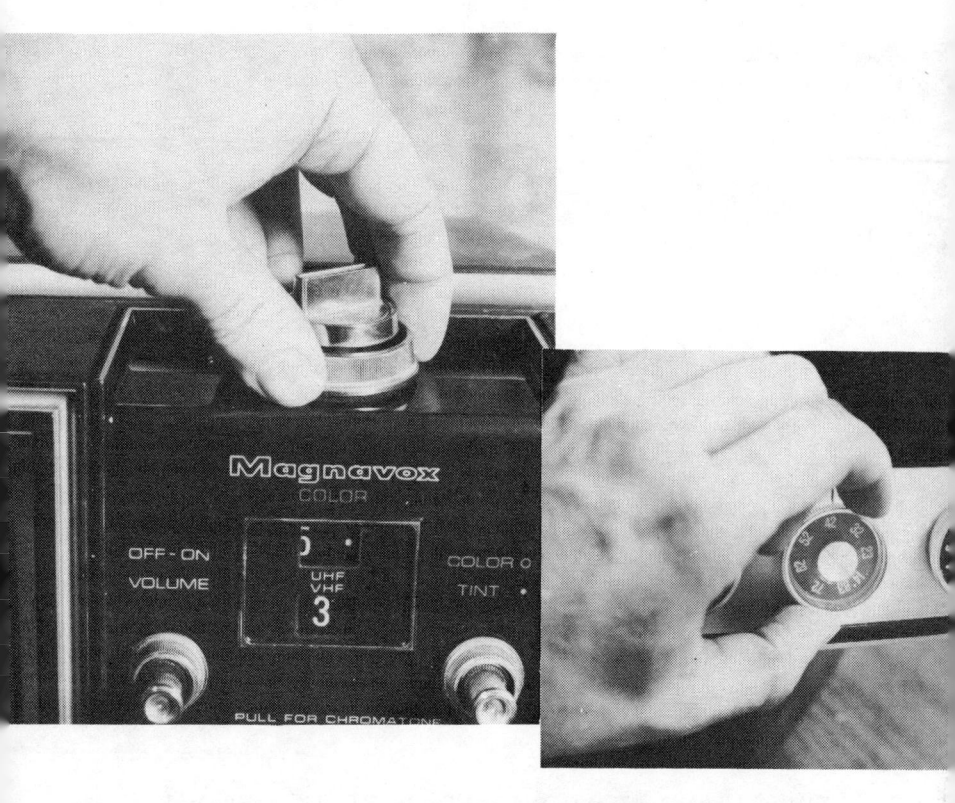

FINE TUNING IS MORE IMPORTANT TO COLOR than most people know or understand. The reason is highly technical but has to do with the way chroma/video/sync is transmitted on the television station signal. Chroma is transmitted toward one side of the station channel. Consequently, the television tuner must very definitely pick up that side of the signal, or color might not come through on the screen. The fact is, you can tune in a black-and-white picture with some color sets and not get color, even though the program is telecast in color.

    The fine tuning knob, behind the vhf channel selector, gives you a means of "centering up" the receiver's tuning so chroma is picked up as well as video/sync and audio.

    You face the same problem when tuning a uhf station. With a continuous tuner like the one shown, you may find the adjustment for color extremely critical. Only if you know just how to go about setting the knob can you get best color, best video, and best sound.

89

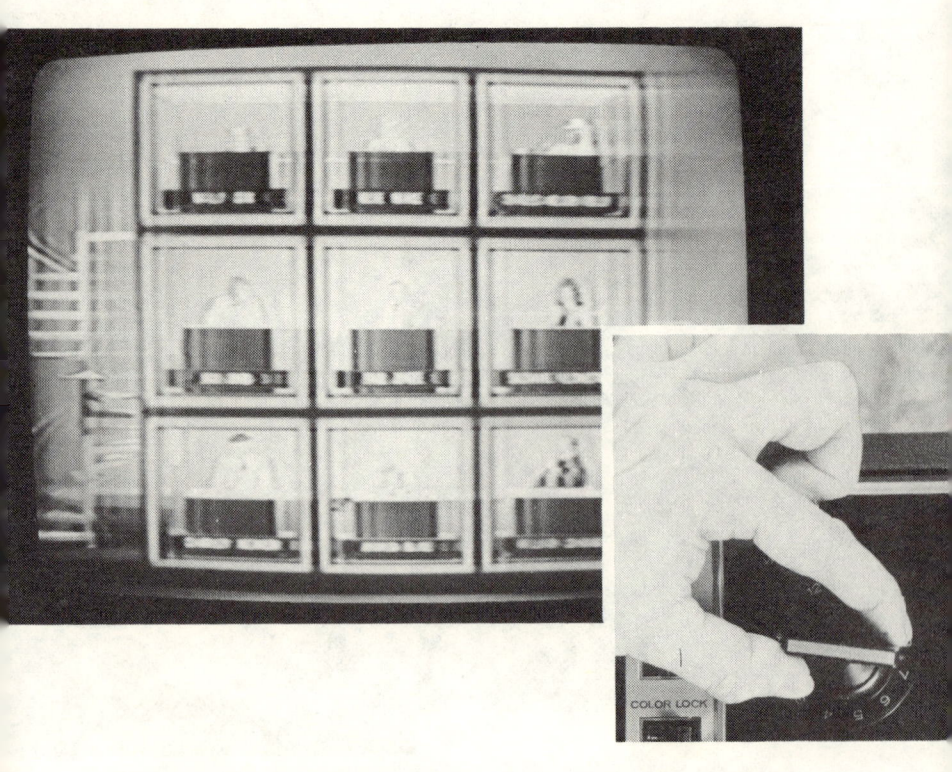

FINE TUNING AT WRONG END OF STATION SIGNAL

The smeared and multiple-image look of this picture on the tv screen makes a dead giveaway to where the fine tuning knob is set. Sound may be poor, too. Neither audio nor video comes through very clear, because the tuner isn't set at a point where they can be amplified by other circuits in the receiver.

Worse, chroma is not picked up at all. This setting of the fine tuning makes no color picture whatever—just the poor black-and-white one you see.

You might mistake the ghosts—the multiple images—for the kind you get when your antenna isn't adequate. A quick twist of the vhf fine-tuning knob or the uhf tuning dial straightens you out on that point. Multiple images caused by improper tuning shift as you turn the knob, even slightly. Antenna ghosts don't shift at all with tuning, however finely you manipulate the dials.

If you see this pattern on your color tv screen, you know you're tuned to the wrong side of the station channel.

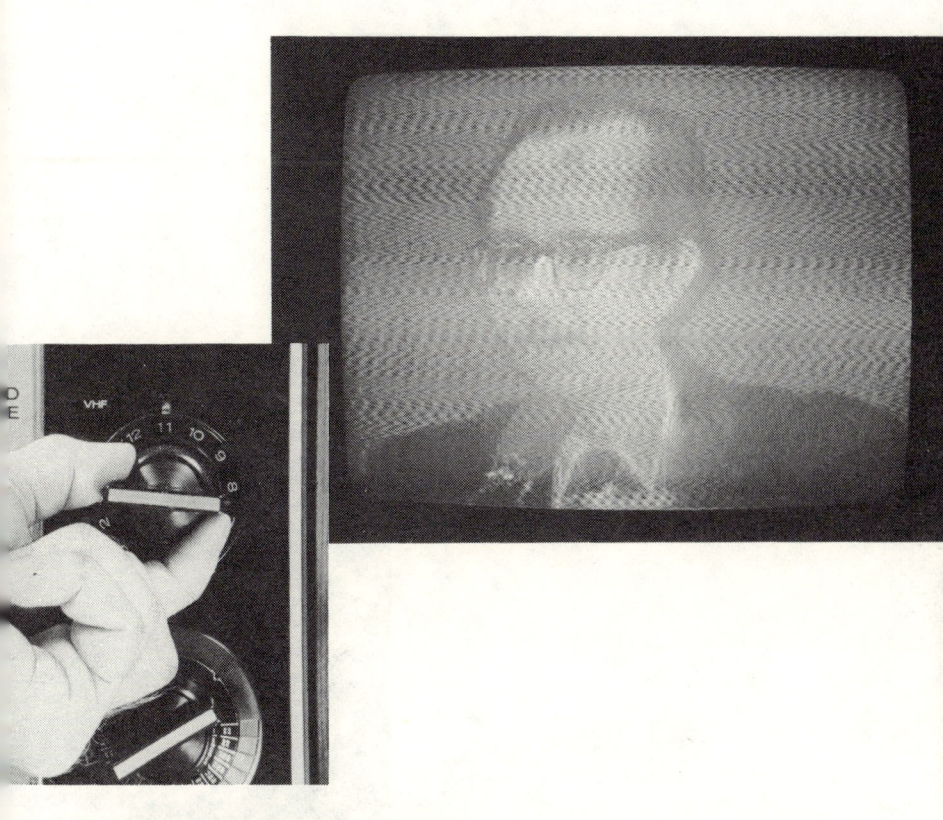

START BY TUNING TO THIS END

This might look like a funny pattern to tell you is "right" for tuning in color. But you'll get your best results if you start with the screen looking like this. You'll be hearing good sound, and the picture might be even harder to see than this. The moiré pattern vibrates with voices you hear in the program.

You've turned the fine tuning to the end that lets the chroma/video/sync and audio signals come through the very best. In fact, you've gone too far. But you need to see this pattern to know which side of the channel signal you're on. Once you've made sure you have the tuner on this side of the station channel, you're ready to tune carefully for the spot that gives best color.

It is NOT important to have Color turned up right now. In fact, you might do better without color visible. What you're doing here is aiming toward getting the tuner adjusted exactly right.

THEN CAREFULLY BRING FINE TUNING CLOSER

The wiry pattern has almost cleared up, but not quite. You want the tuner almost showing the fine moiré, but not showing it. The screen looks like this when you have the tuning almost right.

The moiré pattern tells you that certain signals are still not mixing quite right inside the color set. If you didn't turn Color down before you started tuning, you'd be seeing color at this point. It may not be good color yet, because you have four more major steps to go through. Here, you've just about completed Step 1—fine-tuning the station precisely.

Handle the fine tuning knob delicately from this point, or you'll go too far. Just barely clear out the pattern. Apply equal or greater care to fine tuning with the uhf tuner knob.

THE STATION IS FINE-TUNED PRECISELY

You've completed Step 1 of the five steps to better color when you have the station tuned in correctly. The video/sync, the chroma, and the audio signals are all being picked up properly from the station. This is the only position of the fine tuning that gives you the best color signal the station can send out.

Remember, start with the fine tuning knob in the direction that gives the picture filled with dark, wiry patterns. Then carefully clear the patterns out of the picture, *and turn no further.*

A color set with automatic fine tuning (aft) does the critical part of this tuning for you. Yet, the aft only can handle so much. For best color, you disable the aft and fine tune as best you can by hand. Then let the aft take over and hold the tuning centered accurately for you.

MAKE SURE OF THE HORIZONTAL HOLD

This may seem like an unnecessary step to you, but it's not. Several circuits in the color receiver depend on the horizontal sweep section (that's what sweeps the picture tube beams from side to side in exact synchronization).

The picture above shows what the screen looks like when horizontal circuits have fallen out of synchronization. But they can be slightly off without your seeing this pattern. The way to tell is this: flip the vhf channel selector off station and back very abruptly. The shock to the circuits will throw them out of sync if they're holding only borderline. You can straighten up the picture with the Horizontal Hold control. Twist the station selector more than once, to be sure you have the horizontal control centered for best stability.

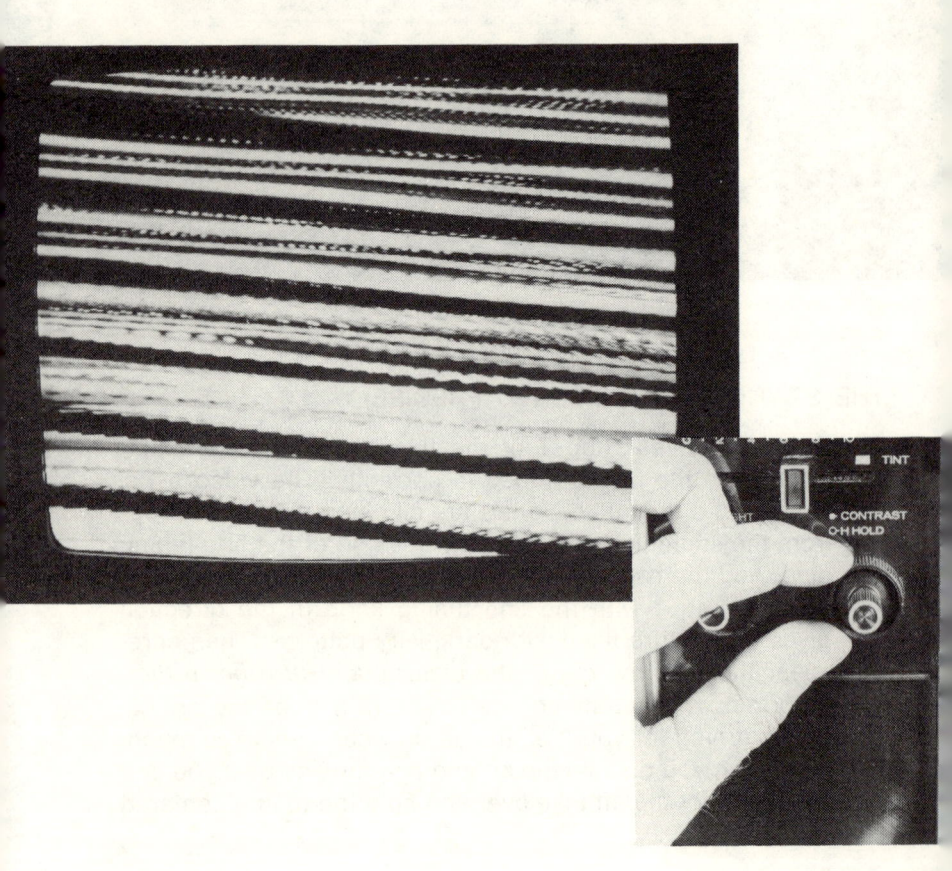

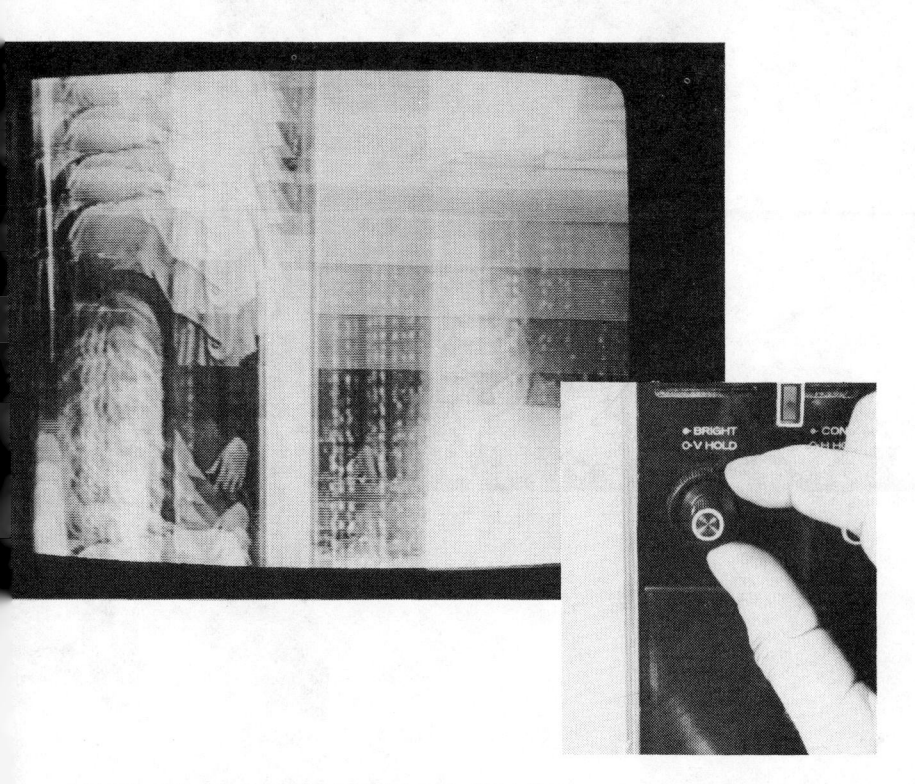

MAKE SURE OF THE VERTICAL HOLD

Again, this may seem unnecessary, but this is an important part of Step 2 in getting the best color picture. The tendency is to skip this, especially if the picture seems to be holding still vertically. But, as with horizontal, certain color circuits depend on the vertical sweeps (which move the picture-tube beams up and down). You can't let the vertical circuits operate even borderline.

Correct procedure for adjusting the Vertical Hold control is as follows:

Turn the Vertical Hold knob in whatever direction (usually counterclockwise) makes the picture roll *downward*. Critically, turn the knob back until that long black bar that moves down with the picture almost stops, preferably across the middle of the screen. Then turn the knob a little more to roll that black bar—and the picture—*upward*. The black bar should lock solidly out of sight at the top of the screen. Turn the knob just a hair further, to lock it even more solidly. But be careful; if you move it too much, the whole picture will start jumping upward.

STOP ANY VERTICAL JITTER

The bouncing in this photograph is exaggerated, but you should be on the watch for any vertical jitter at all. The picture should not jerk up or down (unless by chance it does when your furnace kicks on).

The cure is with the Vertical Hold knob. This may take a delicate touch. If the jitter is too touchy or keeps coming back, you need a repairman.

Start just as you did on the preceding page, by rolling the picture downward. Then move it back up till the black bar locks out of sight at the top. Now watch for jitter, and move the Hold knob ever so slightly to stop any jitter you see. If the set's jittering, you'll see it in the first few seconds after you lock the picture as just described. Sets are not prone to jitter unless you've left the Vertical Hold control set just slightly wrong. Clear it up. That and stabilizing the Horizontal Hold complete Step 2 of your procedure toward the best color picture you can get.

STEP 3: THE BLACK-AND-WHITE PICTURE

You may not understand why an important step in tuning the best color picture lies in making black-and-white look right. But accept the fact that it is necessary. Here's why. Remember back on pages 38 and 39 how the two elements of a color picture are video and chroma? Well, it happens that the video element—the luminance that goes into each portion of a color scene—is just about as important to color as the chroma. The luminance determines just what shade a color takes, how bright or dim it is, and even what color it looks like to your eye. Even more, the luminance signal deeply affects the phosphor dots on the picture-tube screen. Without proper brightness, some of them even glow off color.

So the next step in getting color to look right on the color television is to make monochrome look as it should. That merely means a good range of whites, various shades of gray, and blacks. (With chroma added, those seeming blacks may become some important color.)

Knowing this, turn the Color control all the way down and adjust Brightness and Contrast as described on the next four pages.

TOO BRIGHT won't work. In black-and-white, if you turn up the Brightness control too much, the blacks are all washed out and gray-looking. A color picture with too much light on the screen looks even worse. The raster—and therefore all elements of a scene—may even go out of focus when brightness is too high. The only cure for this is turning down the Brightness knob.

NOT BRIGHT ENOUGH won't do, either. Just about everything in the monochrome picture gets too dark to see. The answer is a little more of the Brightness control. Don't try to adjust the contrast of a picture like this until you have most of the whites turned up to something like normal. A color picture with so little brightness would appear dark. Besides that, some of the colors might be completely off-hue.

TOO MUCH CONTRAST can make a monochrome picture harsh and unpleasant to watch. You lose much of the detail, since there are no intermediate grays. Everything seems stark and raw.

The cure lies in adjusting the Contrast knob. You might have to juggle brightness a little, too, but you won't know until you've reduced Contrast to a viewable level.

Setting contrast by a color picture is nigh impossible. You have just too broad a range of luminance values to deal with. The intensity of various colors can fool you mightily. That's why adjusting Brightness and Contrast is the third step of getting a good color picture, and why you set them with the Color control turned completely off.

WEAK CONTRAST can show its very worst effects in a color picture. Without the luminance values of the video (the Contrast control increases video reaching the picture tube), the colors don't even have proper hues. Some colors seem to disappear altogether.

Yet, as pointed out on the opposite page, you can't judge proper contrast at all when a color picture is on the screen. So, you turn the Color knob completely off while you adjust brightness and contrast.

Here's a procedure for the two that works well. Turn the vhf channel selector away from any station. Turn Brightness and Contrast all the way down. Then bring brightness up until the raster looks fairly bright and normal. Turn the channel selector to a station. Bring up the Contrast knob until the picture looks reasonably dark in the shadows and light in the sunlight or on faces of actors.

You might have to juggle Brightness a little, but don't change it much. You might find yourself trying to make up for too much brightness by turning on too much contrast. The picture will look fuzzy and lack detail. Back the brightness off till things look sharp yet still show good whites among the blacks and grays. Then you're ready to add color back in.

THE FINAL TWO STEPS FOR TUNING THE BEST COLOR involve the two color controls: the Color knob and the Tint knob. You've gone through the first three color-tuning steps, and they're exceedingly important to your final result. The station is tuned precisely, the picture is stable on the screen, and the gray scale has been smoothed by proper settings of the Brightness and Contrast controls.

Adjust these final two controls only in the following sequence. First turn up the Color (you had it at zero for adjusting Contrast and Brightness). Then vary the Tint back and forth to get the most accurate and pleasing skin tones.

The knobs on the set shown here are sliders. Quite a few brands today use slider-type controls. But they function no better—and no worse—than round controls.

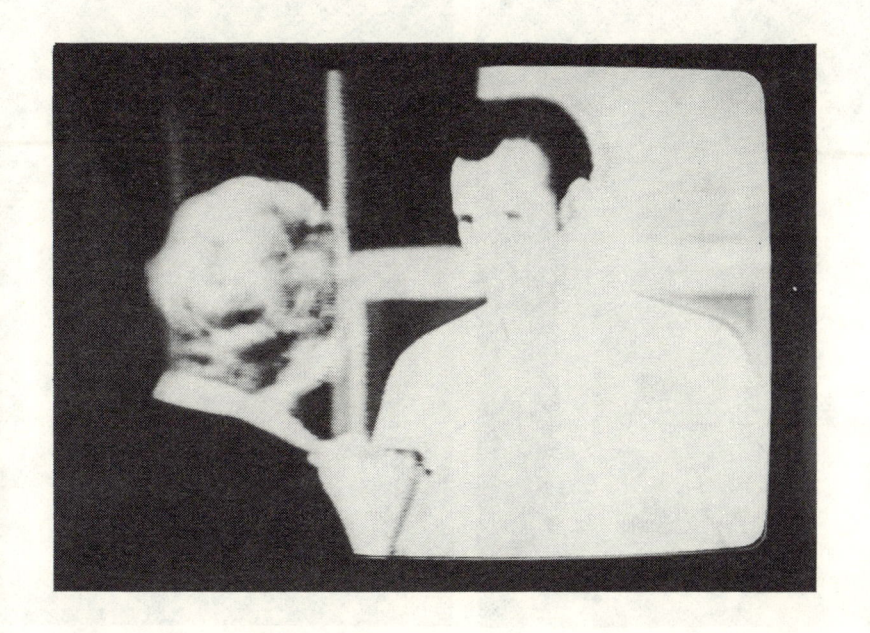

TOO MUCH COLOR is the most common mistake of a newcomer to color television. Use your imagination a little with the picture above, and you can know what too much color looks like. If you have a color set handy and a color program tuned in, just twist the Color knob wide open and you'll get a look at it.

One characteristic of a too-high setting of the Color knob is red overwash in many areas of the screen. What looks white in this photograph is really a highly saturated red-orange coloration that has taken most detail out of the picture.

Notice also the ragged pattern that shows up around any sharp edges in the picture. They are mottling all over the screen, but the overwash covers some of them up. You might see them if you turned brightness down, but that would foul up your monochrome control settings. Don't.

For best color, start with the Color knob at zero. Bring it up until you just barely get color. Don't go much above that until after you've set the Tint knob. Then, and only then, bring up the color to a pleasant viewing level. You may still go for too much color until you get used to watching color tv. Then a light pastel face shade will give an overall best appearance to other colored objects in each scene.

ADJUST TINT FOR FLESH COLOR. This is the final step toward getting the best color picture you can. You make this Tint adjustment with the Color control turned up only a small amount. Otherwise, it's hard to tell when you've really got flesh color in an actor's face or just some shade of orange. You might add a trifle more color afterward, but start out with only a little.

Begin with the Tint knob about centered. Then, move the knob in one direction. The faces will become greenish. The left-hand picture above was taken when color hues were off in the green direction. Use your imagination and notice the shading of the faces, the sweaters (which happened to be red originally and were shifted to blue), and the originally reddish borders on the desks.

Move the knob the other direction and faces become blue. The sweaters above (right) turned sickly green, and the desk borders a shade of blue.

Move the knob to wherever you find flesh color best. You can refine it with just a little more color if you prefer.

AND THAT'S THE TOP-NOTCH PICTURE YOU WANTED FROM YOUR COLOR TELEVISION!

Chapter 8

Fixes You Can Try (and Some You Shouldn't)

The very best color television sooner or later goes *kerplooie*. What you do then depends on many things. The final two chapters of this book will help you make up your mind.

In most cases, your safest set of tools is pictured at the bottom of this page. Call an expert. But if you're a little more adventurous than that and are willing to put a little common sense to work (and some knowledge from this book), turn the page and begin finding out what color tv "repairs" you can handle yourself. The explanations include where to leave off.

AVOID SCENES LIKE THIS. You can get into more trouble pulling the back off a color tv set! For one thing, the set shuts off when you take off the back. If it didn't, dangerous high voltages could send you bouncing off the wall. Even with the set turned off, certain unpredictable voltages are stored in the most unexpected places.

So, rule number one is this: Don't attempt repairs that require you to pull the back off the set. You'll find a fair number you *can* do illustrated on the next several pages. Be satisfied with those. If you know enough about color tv to be poking around safely inside the chassis, you don't need this chapter anyway.

Leave the back on. You'll stay happier and healthier, and so will that intricate mechanism—your color tv.

YOU HAVE TO RECOGNIZE SYMPTOMS. Some of them mean that you need only turn one or more of the operating controls. Now that you have instructions in how to adjust them properly, you can experiment a little.

The picture above, for example, was made with Horizontal and Vertical Hold knobs both turned wrong at the same time. Even from this photo you can't tell what the pattern really looks like on the screen, because it actually wiggles all around.

You can duplicate this messup yourself by way of experiment. Knowing what you did to foul the picture up, you can cure it easily. If it even happens when you're not expecting it, you'll know what to do. The same is true for all the other operating knobs. You can try messing up Color, Tint, Brightness, Contrast, uhf and vhf channel tuners, and any others you see sticking out.

You're not ready for the knobs and shafts in the back yet, but don't be discouraged. Your first job with them is to learn what symptoms in the picture mean one of them is misadjusted. Then, if your effort doesn't cure the symptom, you know you've done what you can and should call a technician in to complete whatever repair is necessary.

STUDY THE SYMPTOMS IN THE PICTURE. For example, what do you see wrong with this television picture? Examine it all over. Little things count.

The top is squashed down. See how much shorter the top boxes are than the center and bottom ones? You could see the same symptom with people standing in the scene. Their heads and shoulders would be smashed down, and their legs and bodies would look too tall for the rest of them.

This is caused by a trouble in the vertical sweep section of the tv set. For some reason, the picture-tube beams are not being swept as fast near the top as they are at the middle and bottom.

A technician would say "vertical linearity is wrong." And the first thing he'd do, you can try, too. He'd attempt to adjust the Vertical Linearity control. It takes a screwdriver. If you can find the control (labeled V LIN, sometimes), turn it a little bit each way to see if the top of the picture pulls up where it belongs. If not, consult the next page before you give up.

A SIMILAR MALADY SHRINKS TOP AND BOTTOM. Another adjustment, the Vertical Height, makes up for small circuit changes that occur. Such changes over a period of time are normal, which is why the adjustments are there. The Height control (often labeled HT or V SIZE) usually is beside the Vertical Linearity adjustment. They operate in associated circuits inside the chassis.

Notice that the height adjustment appears to affect both top and bottom of the picture. In the photo above, the tv picture has raised up from the bottom and come down from the top. The tubes or transistors that sweep the picture tube beams up and down may have grown weak. The control lets you restore the right height without expensive repairs—at least not yet. You might have to have the trouble repaired later when it gets worse, but not before it's necessary.

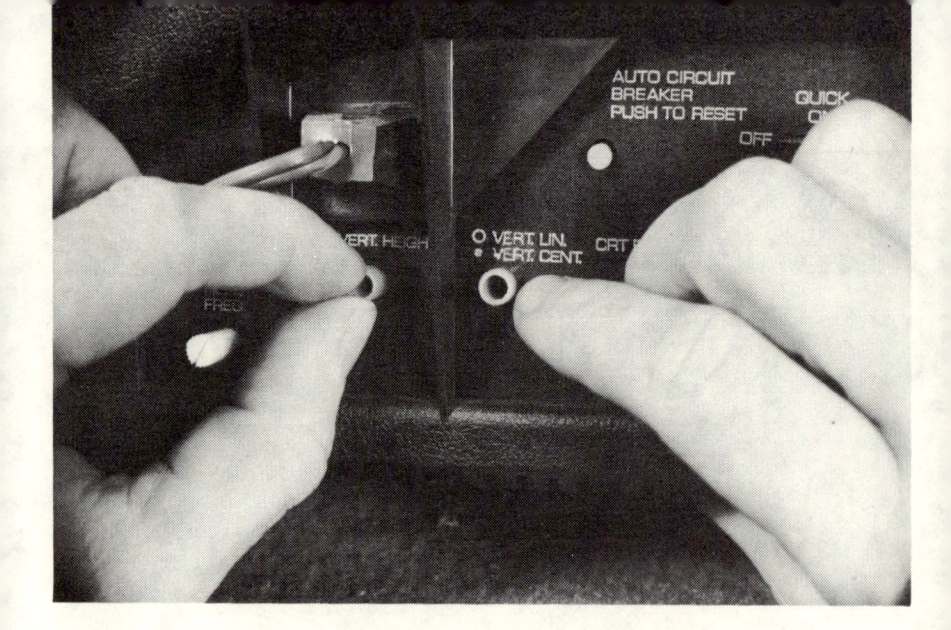

HOW DO YOU FIND THE ADJUSTMENTS? Merely by looking for them. Try inside the soapdish, or along the rear of the set. They're always labeled, and you can reach them with a screwdriver. Some, you can turn with your fingers, like those shown in the photo above.

Some Height and Linearity controls are inside the chassis, like the ones shown below. Leave them alone when you can't reach them without removing the back of the color set. You could mess up more than you help.

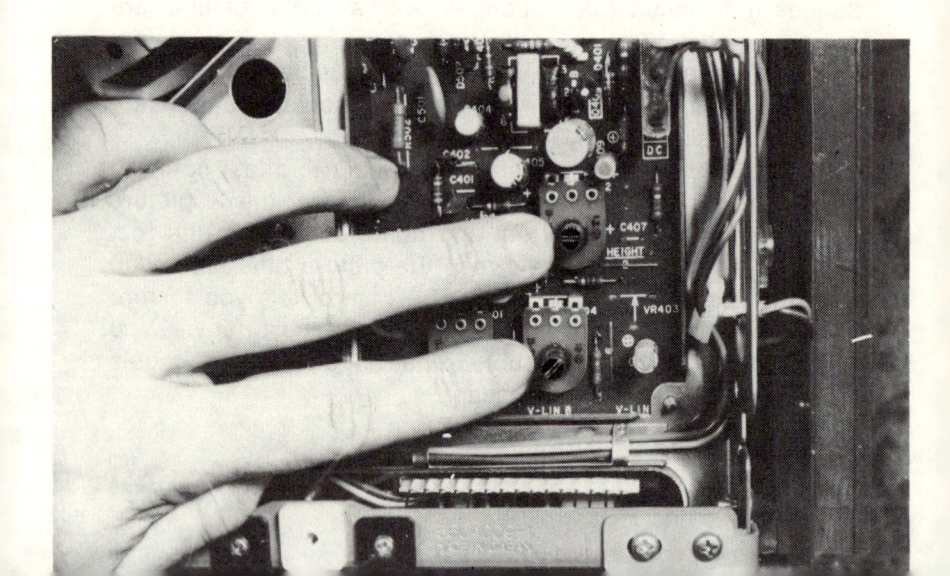

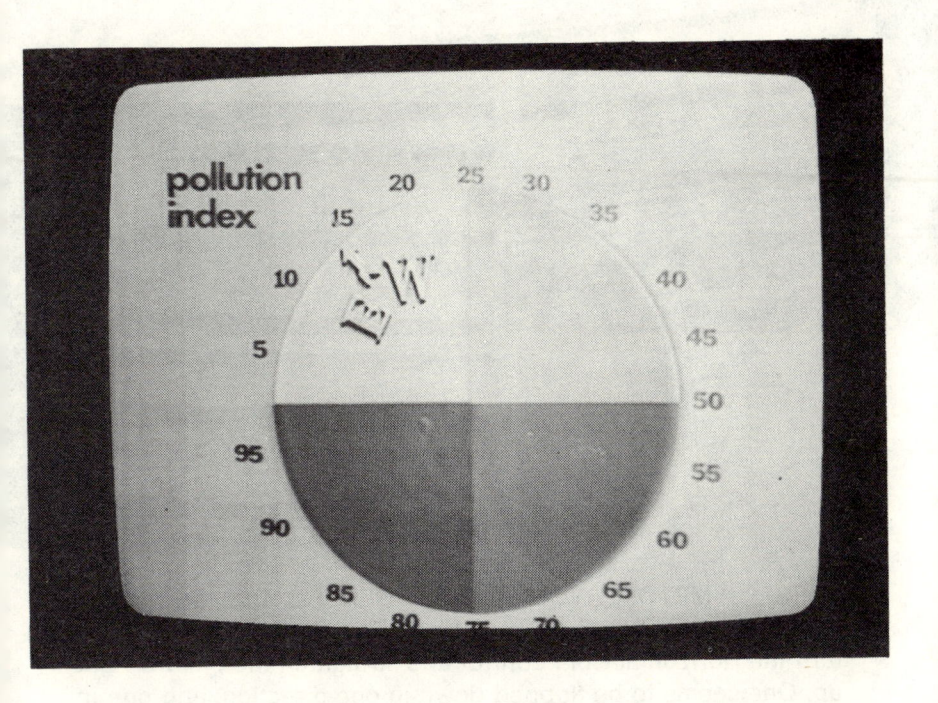

USE VERTICAL LINEARITY AND HEIGHT TOGETHER to make the picture look right from top to bottom. If you can view a circle while you're adjusting, or when you've got the settings nearly right, by all means do. It's a big help, as you'll see.

First turn both adjustments so the picture is as thin across the middle of the picture tube as it will go. Turn the Linearity just enough to pull the top up a little bit. Then turn Height a little to bring the bottom down about the same amount. The circle, if you have one on the screen, will stay squat but not egg-shaped.

Then turn up the Linearity a little more, for the top. And then the same with Height, for the bottom. Keep alternating them until the picture stretches evenly from top to bottom.

The circle should end up perfectly round. If it's too short, you need a little more adjustment on both shafts. If it's too tall, you've gone too far. Egg-shaped with a flat top means not enough Linearity adjustment; with a stretched top, too much Linearity. With flat bottom, the circle says Height isn't pulled out enough; with stretched bottom, Height has gone too far. Make the circle round.

THESE SYMPTOMS YOU'VE SEEN BEFORE. But the remedy you think should fix them might not. They look as if you could turn the Horizontal Hold control and straighten the picture right up. One seems to be flopped down in one direction, and one in the other. Two factors can foul you up here.

(1) The Horizontal Hold control may not be readily accessible. On many color receivers, particularly portables, you find the H Hold knob protruding from somewhere on the side or back of the cabinet. If the cabinet is dark, the shaft is usually white, but it seldom has a knob on it. You'd take it for one of those "leave-it-alone" adjustments.

(2) The Horizontal Hold control may be labeled something else. The most usual is Horizontal Frequency or some abbreviation of that. In these instances, you can expect it to be hidden away in back or inside a soapdish. Whatever the label, its use is the same.

One caution: Some H Freq shafts don't have stops. You can turn and turn and turn, on and on. If you happen to start out in the wrong direction, you could go on forever—eventually blanking out the picture altogether. With one of these, if a few turns one way don't straighten up the picture or make the ropey lines larger and fewer, back up and go the other direction. If eight to ten turns either way seem to have no effect, give up and call your service technician.

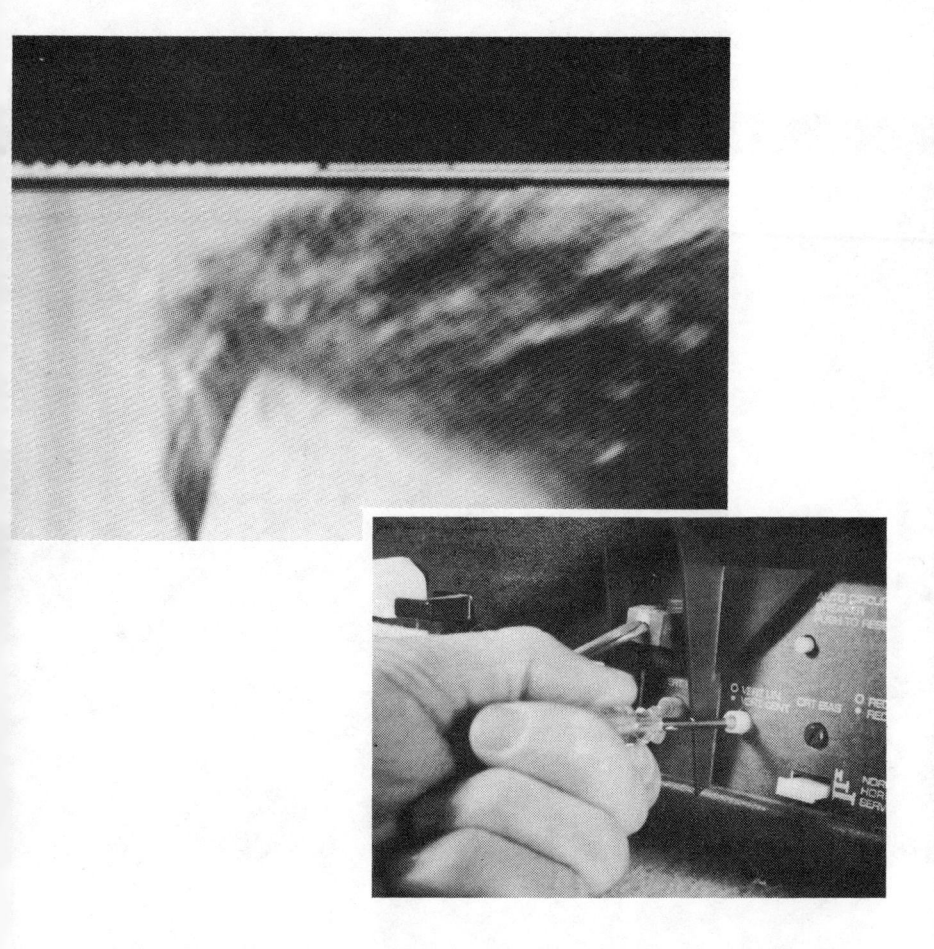

SOMETIMES THE PICTURE SLIPS DOWN FROM THE TOP, or up from the bottom, without actually changing size. If it's down from the top, and a color program is on, you can see the white patterns that are revealed in the picture above. The slipping is a result of Vertical Centering becoming a bit off-center. (By now you must have noted that anything affecting the up-and-down is labeled vertical-something.)

Not all color sets put Vertical Centering where you can reach it. If you can get at them with a screwdriver, be sure you've got Vertical Linearity and Height okay before you mess with Centering. Give Height and Linearity the circle test described on page 111. Then center the picture so it overscans a bit at both top and bottom.

Horizontal Centering works the same way if the set has such a control and if you can reach it. On most sets, you can't.

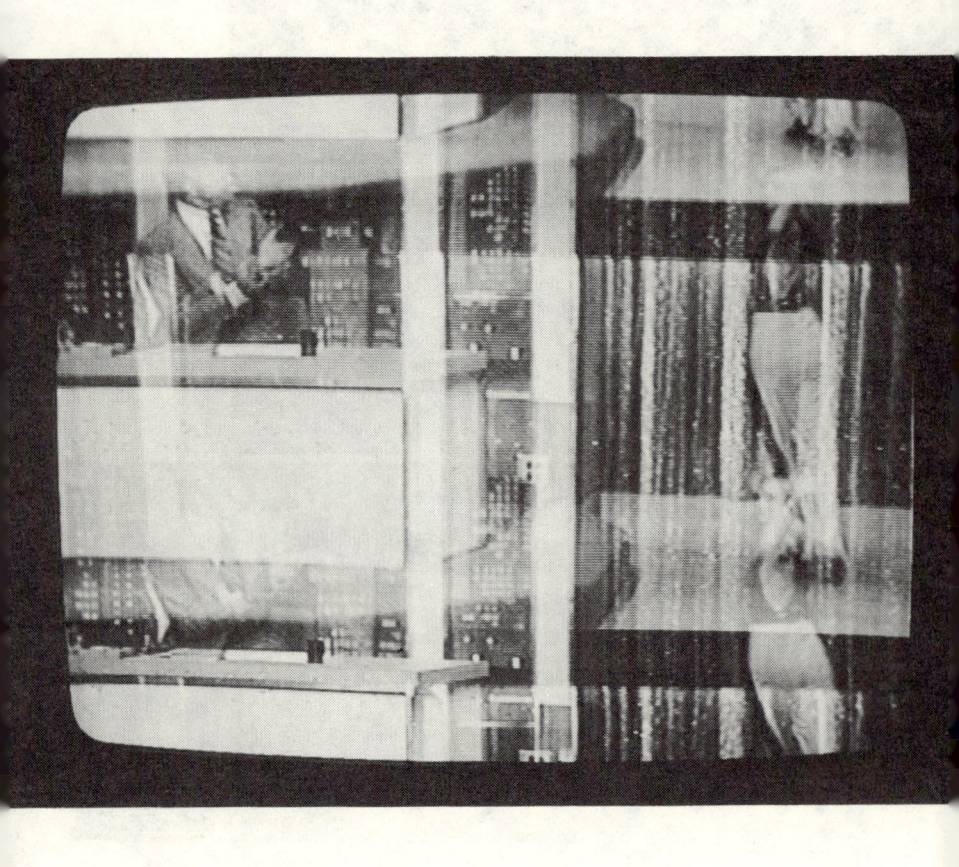

VERTICAL ROLLING TOO FAST can usually be stopped by the Vertical Hold knob. You can tell if the knob is working by whether it slows down the rolling or not. Or perhaps it changes the direction from up to down. If so, the Hold knob is doing what it's supposed to, although if the picture doesn't lock you might think it's not.

Remember the sync signal that accompanies chroma and video from the television transmitter? One part of that sync is supposed to lock in the picture and keep it from rolling up or down. If it does, everything is okay unless you turn the Hold control too far from where it should be. But with the Hold set properly, the picture should stay still.

IF VERTICAL HOLD ONLY SLOWS DOWN the picture and leaves it floating, but can't lock it into place, then the sync signal is definitely missing. A color tv technician would say "the set has no vertical sync." And if that happens to your color set, you'd better call him in. It takes expertise to find out what has blocked vertical sync from locking in a picture for you.

SPLOTCHES OF COLOR ON THE TV SCREEN have already been described. They come in pink, lavender, green—just about any color. Usually they are caused by one of two things. A metal screen you can't see inside the face of the picture tube has become slightly magnetized, or the deflection yoke and purity rings have slipped (referred to on page 84).

You can't do anything to cure either one. They're a job for your service expert. But you can take some steps to reduce the likelihood of their occurrence.

Keep any large metal—particularly iron and steel—objects away from the front of the tv screen. Most especially, don't let anyone near the set with magnets. Don't use your flashlight close by; it can create a magnetic field that leaves colored streaks. Keep electric trains away. If you run the vacuum cleaner nearby, move it several feet away before you turn it off. Don't put Christmas-tree lights too near, or any kind of flashing electrical device. Even certain electric clocks have been known to leave discolored streaks if they are kept a mere few inches from the tv screen.

Look at the screen with Color turned down. If there are hazy areas of color, phone the technician. He'll probably want to bring his degaussing coil (page 83), so tell him what you think is the trouble.

WHOLE PICTURE HAS GREEN (OR BLUE, OR RED) CAST.
You can see this fault more on black-and-white pictures, as a rule. If you watch your set every day, the odd-color shading may develop so gradually you won't even notice it. Someone else may mention it. Or, you may watch a black-and-white television some afternoon and come home to wonder why yours looks so greenish or bluish or whatever.

Or . . . you may just suddenly find yourself some evening wondering why all the shadows in your color tv picture look green or blue. Or why things you know are white have been looking pink or purple.

Turn the Color knob down, to leave a monochrome picture. Look at a white sheet of paper and then back at your tv screen. An unseemly overall cast to the picture, or to a blank raster with no station, signifies what your technician will call a "gray-scale problem."

This is something you can perhaps do something about yourself. The next four pages explain how.

START OUT WITH CONTRAST DOWN AND BRIGHTNESS UP just a little. Don't overdo the brightness. You want the raster a bit dim. Be sure you turn the Color knob down so you have nothing but a monochrome picture. You can do this adjustment with no picture at all. Just tune to some empty channel.

   Your intention will be to even up the mixture of three colors that make the raster on the screen. That is, you want just the right amount of red, green, and blue to make a perfectly white raster. You adjust them with the three color adjustments labeled Screen. The shafts are at the rear of most sets. If they're inside, don't mess with them. Only try adjusting controls you can reach without taking the back off the receiver.

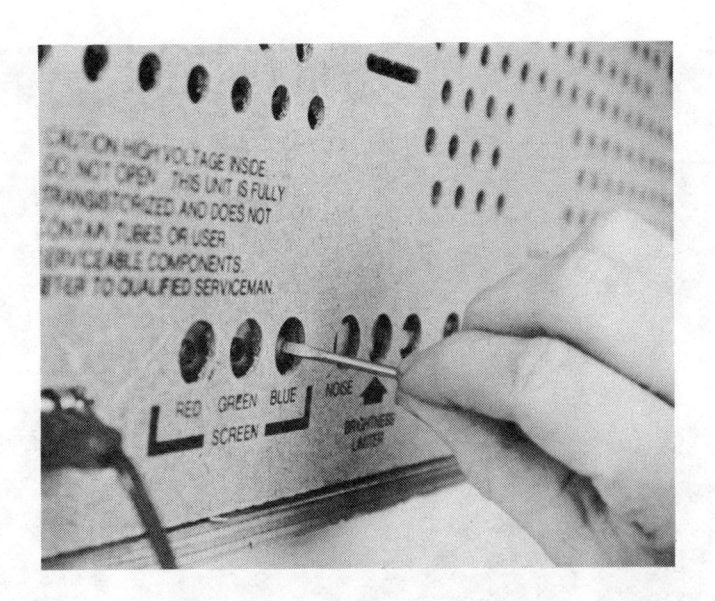

THESE ARE THE SCREEN ADJUSTMENTS that you turn to even up colors in the raster. But you can't turn them haphazardly. There's a very orderly procedure. Follow it, and you'll end up with a white screen. Jump around, and you'll be lucky if you get the raster even close to white.

1. Turn all three adjustments down completely, counterclockwise. The tv screen probably goes dark.
2. Turn up Red Screen till you see a noticeable red raster all over the screen. Don't raise it too far, just enough that the tv screen is definitely red.
3. Turn up Green Screen until the raster turns yellow. You need a good eye for color. Don't settle for orange, and don't go too far green. A lemon yellow, even though it may not look too bright, is what you're after.
4. Turn up Blue Screen to make the raster white. It may look gray, but it should not look bluish or reddish or greenish, or any intermediate shade.
5. Refine whichever one needs to be touched up for perfect white (gray). Hold a sheet of white paper nearby for reference if you need to. If necessary, back all three off and repeat. If you get the yellow exactly right in Step 3, blue will make perfect white.

THAT'S NOT ALL THERE IS TO GRAY SCALE adjustment. On the preceding two pages, you merely set the "dim" end of the gray scale for best white (or gray). Next, you line up the bright end—*highlight* end, your technician would call it.

You do this with Contrast and Brightness turned back up to normal. The adjustments you use are labeled Drive. If the Screen adjustments are easy to reach, chances are the Drive adjustments are, too. If they're inside the back cover, though, you'd still be best advised to leave them alone and phone a trained expert. The inside of a color tv is an unsafe—and sometimes expensive—place to be monkeying around.

FOLLOW THESE STEPS TO MAKE HIGHLIGHTS "TRACK" on the gray scale. You've already (on pages 118–119) set the lowlight end of the gray scale for best white. Now you set the highlight end so it produces best white, too.

1. Bring Contrast and Brightness up for a normal picture. Keep Color turned all the way off so the picture is only monochrome. You can't make these adjustments with color programs showing.
2. Turn all three Drive controls down—counterclockwise. Some tv sets have only Blue Drive and Green Drive. If so, skip the next step.
3. Watching the brightest portions of the scene on the picture tube, turn up Red Drive until you just barely see a pinkish tinge. The less the better, as long as you can tell from the picture highlights that you've turned up the control some.
4. Turn up the Green Drive till you see the pinkish tinge in the highlights change toward orange. Stop there for now.
5. Turn up Blue Drive till the highlights turn just faintly bluish. If green is up fairly well, highlights may turn a sort of seagreen.
6. Refine Blue and Green Drives—leaving Red alone—till the whole picture looks black and white, without any faint green or blue shading.

You can check *tracking* by turning the Brightness knob up and down. The raster (or picture) should stay white from low brightness to high. If not, repeat these four pages.

AN OUT-OF-FOCUS PICTURE is something you may be able to cure. As usual, it depends on whether the Focus knob is brought out where you can reach it without opening up the back.

If you look carefully at the raster or a picture on the picture tube screen, you can see a large number of horizontal lines. Nearly 500 of them are visible, one below the other, stretching from the top of the picture tube to the bottom. They are called the *raster lines*. And they determine how well focused the tv picture looks to you from a normal viewing distance. (If you get too close, any of the lines look fuzzy.)

The thinner the lines are and the more clearly you can see them when you're up close, the better and sharper the television picture looks when you're watching from back where you should. If the picture and the lines look fuzzy from a distance of four feet or so, you can try focusing.

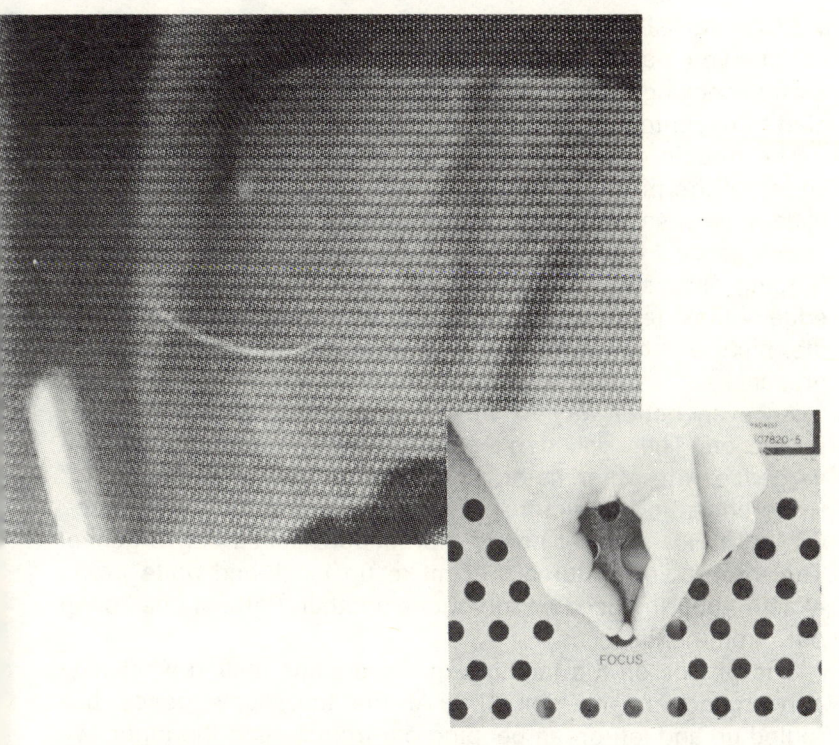

FOCUSING THE RASTER LINES is something you can do easier than you can read about it. Just follow certain rules.

1. Turn Color off and set Brightness and Contrast for a normal picture. Especially set Brightness no higher than you normally watch it, for it has some effect on focus.
2. Position yourself where you can watch the screen and reach the Focus knob, too. You may have to use a mirror.
3. Twist the Focus adjustment back and forth, watching for refinement of the raster lines. Don't look at picture elements; they'll fool you. Move the knob back and forth through the "best" position two or three times just to be sure you have it right.

Some Focus adjustments can be turned several turns; others are like other controls and stop at either end before you go one full revolution. Be careful and don't go too far or force the multi-turn kind; you'll wind up needing a service call.

MISCONVERGENCE is a word you'll want to know about, but not one you may be able to do much about. Misconvergence is a very annoying malady when it gets bad, and you'll probably be glad to pay a technician to restore the set to normal.

The trouble generally shows up first in corners and toward edges of the picture tube. You'll see it as fringing of colors— usually very sharp. Don't mistake ghost-like fringing for misconvergence; that's a different happening. Misconvergence fringing always shows up as color peeping out around sharp edges of images in a scene. Without sharply outlined figures in the picture, you don't see misconvergence even when it's present.

An even better way to view misconvergence is with Color turned off. Then you don't confuse color caused by misconvergence with other color. Look along sharp black edges or white edges in the picture. Is a red fringe bordering it? Blue? Green? If so, your set needs a procedure called convergence. The three color pictures that make up black-and-white aren't exactly superimposed on top of one another. Parts of one "peep out" from behind.

The photos on the facing page give some inkling what misconvergence might look like. At the top, convergence has fouled up and left green peeping out from behind the other two colors. This occurred only in the upper left corner of the screen; other sectors of the screen were okay.

The middle photo shows the same sector of the screen with the misconvergence corrected. Colors are all properly "together" and the picture is pure black-and-white.

The bottom photo shows the screen when a color tv technician has hooked up his test instrument, called a *color-dot-bar generator,* to allow him to adjust convergence. Green lines are showing from behind what would have been white lines. The crosshatch of lines his generator puts on the tv screen shows him, if he interprets the pattern correctly, exactly which of several adjustments he should twist to return convergence to normal.

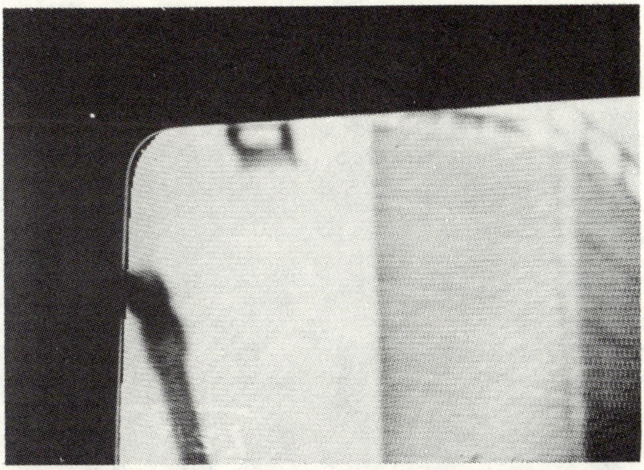

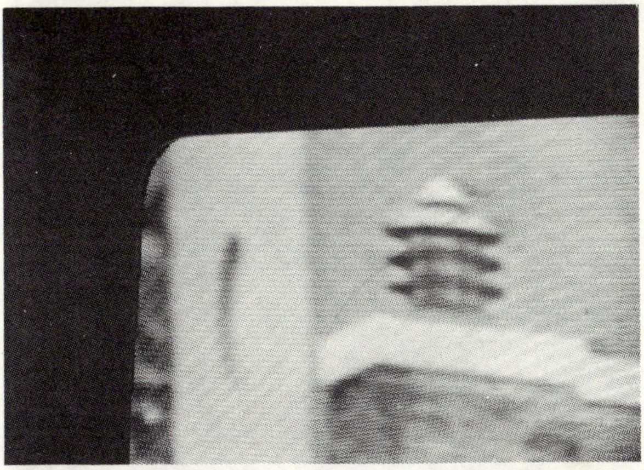

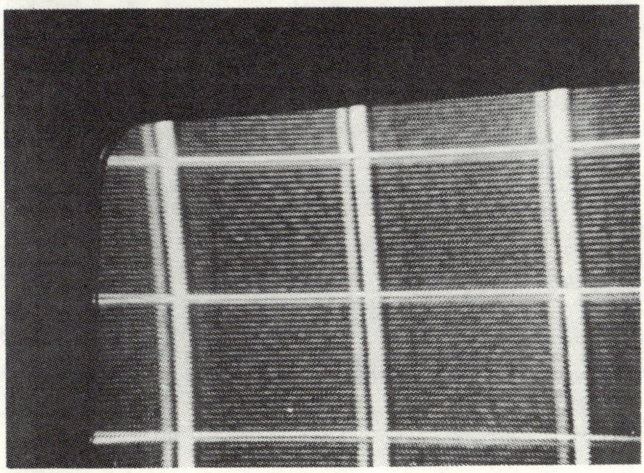

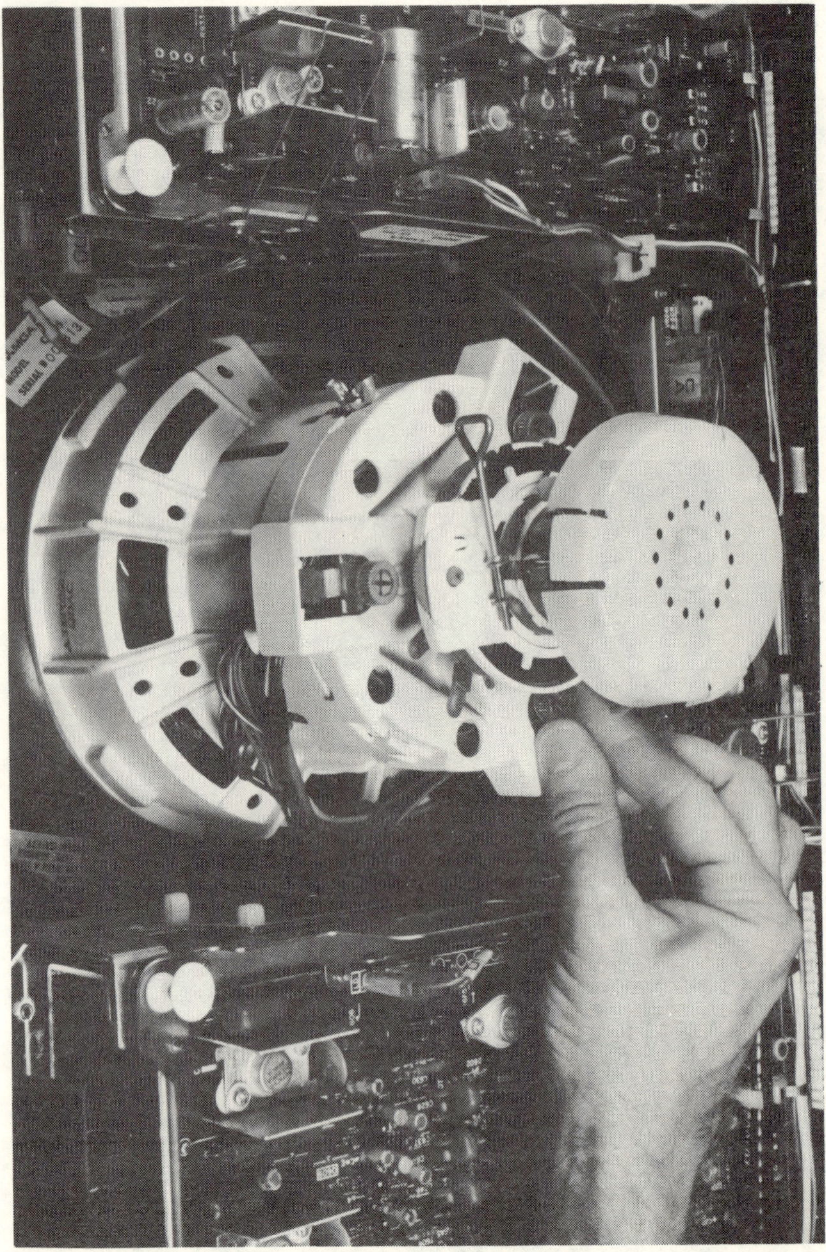

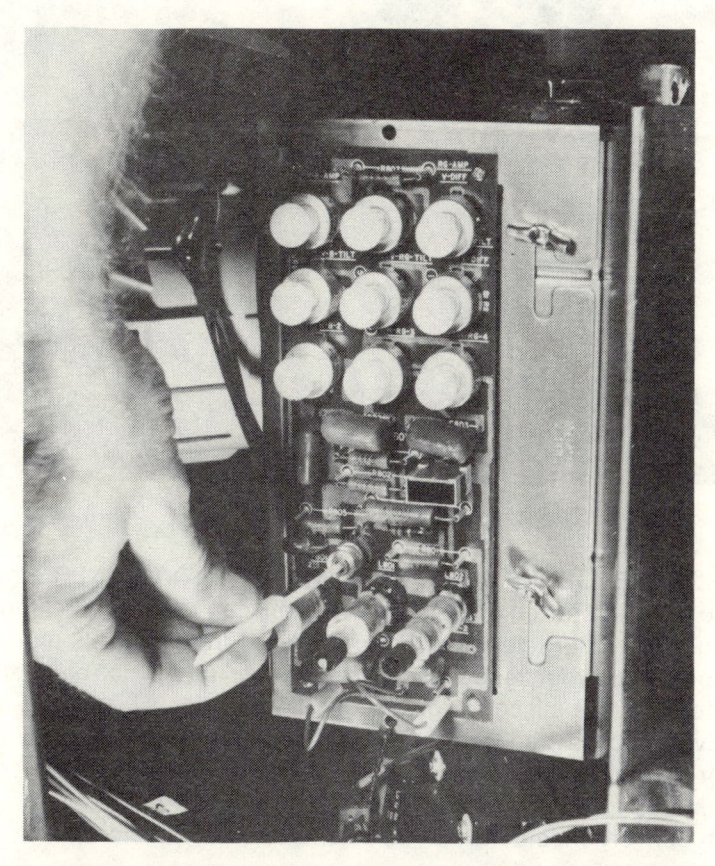

THIS IS WHY YOU SHOULD NOT ATTEMPT your own convergence. For one thing, the controls and adjustments are all inside the back. You shouldn't have it open.

But even more than that, converging the pictures on a modern color television receiver is a complicated and technical operation. The photo at left shows four major adjustments on the convergence yoke that mounts on the picture-tube neck. And the convergence panel illustrated below contains thirteen more. They interact among themselves. Even some experienced technicians have trouble going through the procedure without having to backtrack.

If you ever open up the set and mess with these, odds are all against your doing the picture any good. More likely, you'll necessitate a costly and time-consuming job for your service technician. The rule here: LEAVE IT ALONE.

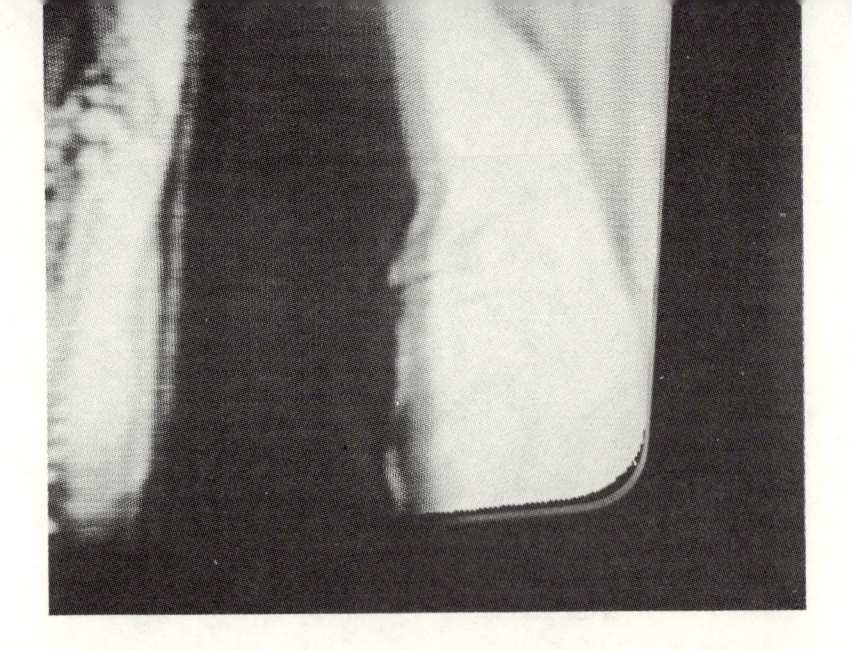

MISCONVERGENCE AT THE LOWER RIGHT CORNER, this time of all three colors. As before, you can't tell it so clearly from the regular television picture, even if the picture is monochrome. But your technician can connect up his dot-bar generator and this misconvergence becomes suddenly obvious.

Even such a badly misconverged screen might be fixed in only a few minutes, provided no one has dabbled with those adjustments (preceding page). But inexperienced hands messing with convergence controls are bound to cost you money.

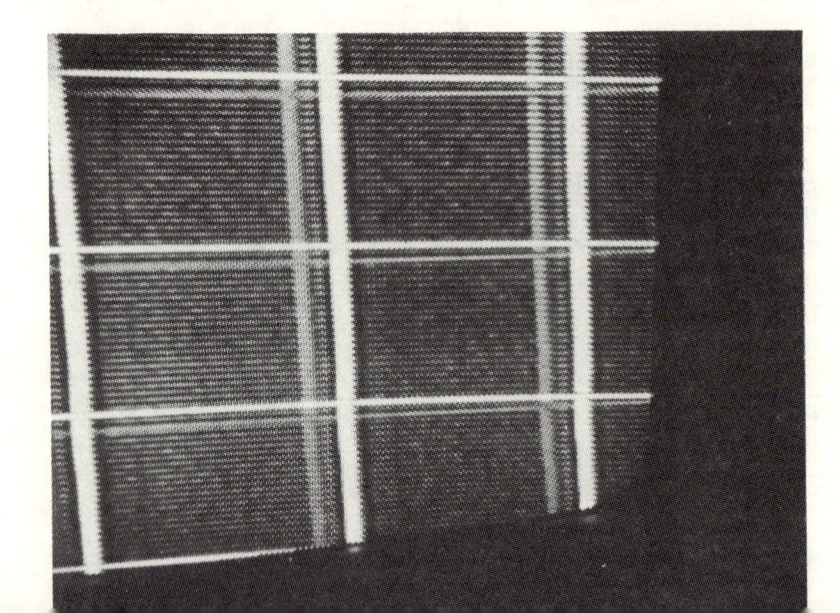

DO NOT TAMPER WITH THIS CONTROL under any circumstances. You should never undertake to adjust the high voltage in a color receiver. There are several reasons.

High voltage makes the picture tube work properly. If for some reason the voltage gets turned too high, the picture tube and sometimes other circuits can give off X rays that might prove mildly harmful. Furthermore, the picture tube can't do its job right if its high voltage is either too high or too low. The wrong high voltage could shorten crt life—and that's costly.

The technician who adjusts high voltage can do it properly only with a special instrument. It may be a long (10 inches or so) pointed probe with a meter attached. It may be a meter he clips into some circuits under the chassis. Either way, that's the only way anyone can tell if high voltage has been set for the right value. Don't let anyone tamper with the High Voltage adjustment, and don't you.

GHOSTS, whether you believe in them or not, can make a mockery of your color television watching. These are not the kind of multiple images you clear up by turning the fine tuning or the uhf knob. Ghosts like the ones shown on these two pages come from multiple reflections of station signal picked up by your antenna.

Sometimes ghosts are caused by a poor antenna, sometimes by the lack of any outside antenna. You'll see them a lot when you try to get by with an indoor antenna. If you live on a busy street, cars will make the ghost images move. Passing aircraft cause flutter in ghostly pictures. Just walking around the room can shift the ghosts picked up by an inside antenna, especially if the station you're watching is uhf.

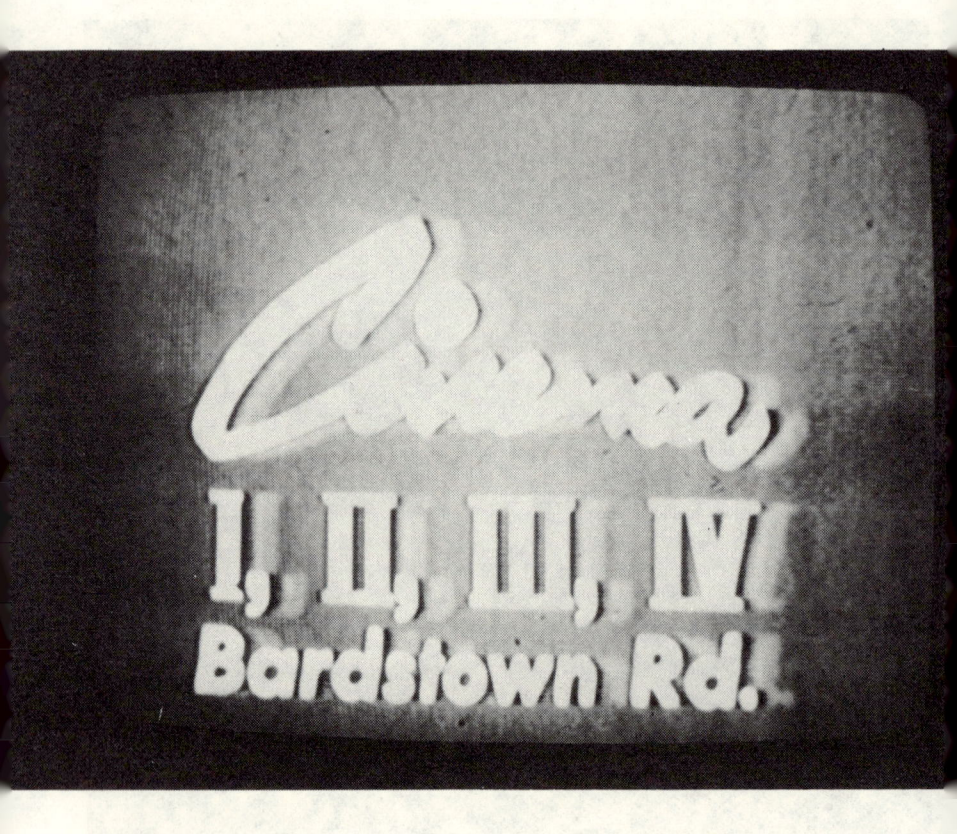

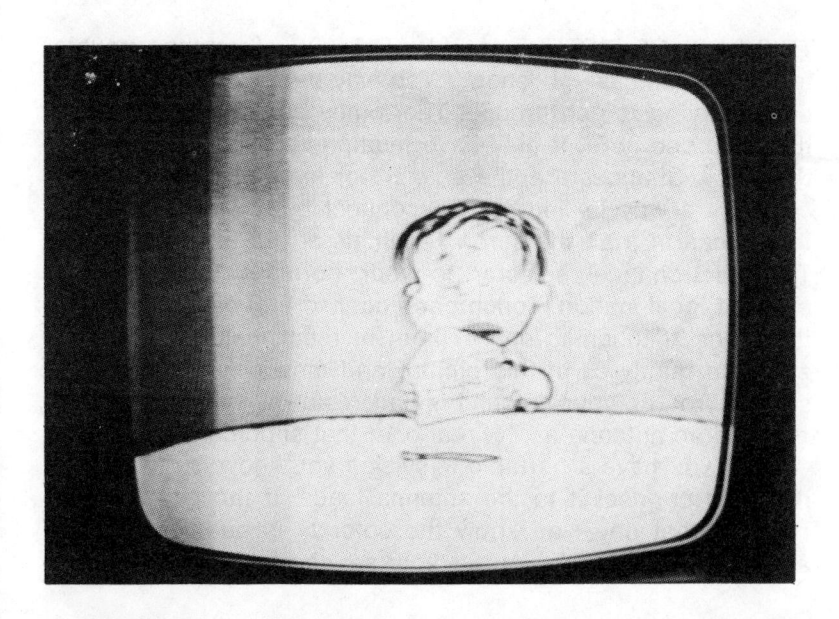

A GHOST THAT TRAILS THE MAIN IMAGE as far as this one does may be a real problem. It can be picked up by an outdoor antenna as likely as by an indoor type. As a rule, the ghost comes by reflection from some building or mountain a few miles away from your receiving antenna.

In this picture, one image of the ghost signal is several inches to the right of the main picture image. There's another even further to the right, which you can't even see. The dark bars along the left side of the screen are the frames of the "ghost" pictures. You don't see the frame of the main picture because it's hidden off the left of the picture tube, as it should be.

Curing ghosts almost always calls for the expertise of an antenna man. He knows the topography of your town and the nature of the signals from your stations. If the ghost can be eliminated, he'll know how. But it may cost you money for a special antenna or for special mounting. Decide whether the cleaned-up picture is worth paying for.

Or, you can get up on the roof and try moving the antenna around to different locations or pointing it in different directions. Most of all, be sure the antenna lead-in has no breaks in it, isn't rubbing metal somewhere, and is solidly connected to the antenna and to the set (or to any couplers you have between antenna and receiver).

THIS KIND OF PICTURE MAY WARN YOU of antenna trouble, too. It's a bad case of "snow." You may be able to detect some picture trying to get through. There may be even more picture than you see here. It may have multiple ghosts along with all the snow. Chances are, the antenna or lead-in is bad.

Here's a way to find out. Disconnect both wires of the antenna lead-in from the screw terminals at the back of the set. Turn the channel selector to your normally strongest and clearest local station. Touch one finger to first one screw, then the other. If touching one or both of them makes the picture show up faintly as in this picture and maybe lets some noisy sound through, your antenna is bad. Your own body is making as good an antenna as the real one; that shouldn't be.

Or, if you have a portable television you know is okay, bring it in and connect it to the antenna leads. If the picture is all right on that one, you know the color tv is bad and not the antenna. Fixing it is a job for your service technician.

IF YOU SEE THIS PATTERN worming its way over your television picture, you're experiencing some kind of radio-frequency (rf) interference. It could be from a Citizens-Band (CB), amateur (ham), or police-radio transmitter. Or, in rare instances, it might be coming from another tv set somewhere close.

You can distinguish between the two. Interference from communication transmitters goes on and off. Your technician would say you have "intermittent or keyed interference." This is a rare trouble in modern color receivers. If you do experience this kind of trouble, and it's bad enough to prevent your watching color television, have your technician help you find the source.

The trouble might be with a transmitter, or it might be with your set. Try another color set connected up right beside yours, if you can; if they both pick it up, that'll prove the fault lies outside the receiver. If it does, then curing it is the responsibility of whoever is causing it.

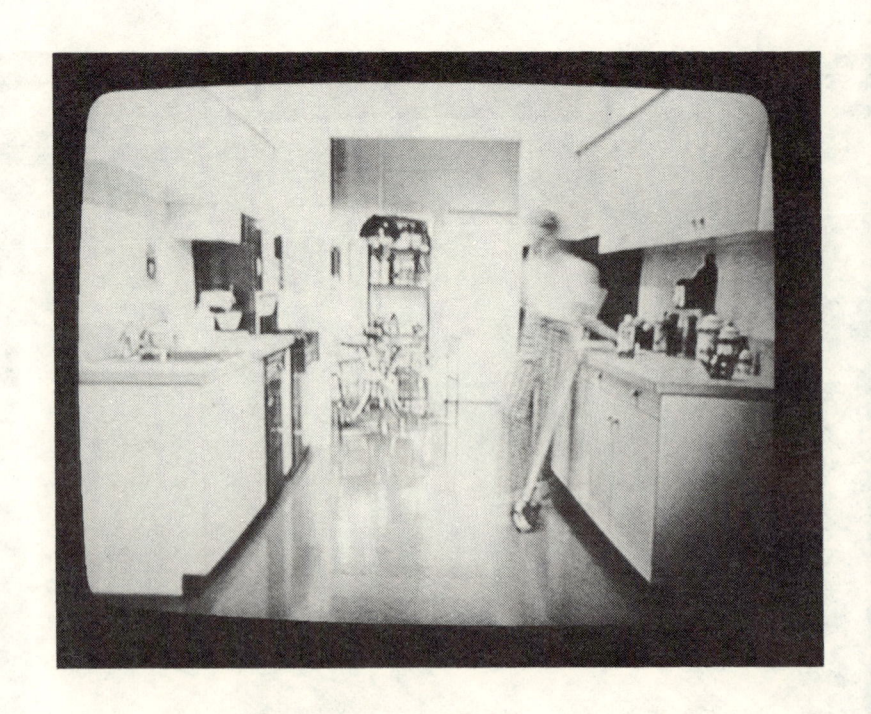

THE VERY FINE MOIRÉ you see in this picture usually can be cleared up by your fine-tuning knob or by tuning the uhf station more carefully. But sometimes not. You might tune far enough to get rid of that overall pattern of wiry lines and lose color.

If critical tuning won't take care of it, or tuning is so critical you can't hold a color signal, phone your color tv repair guy. He'll probably have to do a complete alignment of the signal circuits inside the receiver. It's complex, time-consuming, and costly. He'll charge you well for it. But a thorough alignment on a set that's been running faithfully two or three years may give it new life you'd forgotten it had.

A caution: Pick a truly competent technician. Be sure he's got television alignment equipment, and hope that he knows how to use it. This is about the trickiest kind of adjustment in a color tv. You don't want it botched up by a beginner or someone who has very little bench-repair experience. (It has to be taken to the shop and put on the bench where a number of special instruments are hooked up to it while the technician completes the circuit alignment.)

Chapter 9

Trimming Color TV Repair Costs

Some of the advice in this chapter may sound trite. You may have heard bits of it before. Pay heed to these words nevertheless. Every single idea herein can help keep money in your pocket. Owning and maintaining a color tv need not be exorbitantly expensive, but it can be if you fall into the traps some color-set owners do.

If there's any single admonition that deserves repeating over and over, it's this: Don't entrust your complicated and intricate color television set to someone who doesn't know how to repair it intelligently. That goes for your cousin, your neighbor, and—lest you forget—you.

Training and experience contribute to making a top-grade color tv technician. Still, that's not all it takes. A fellow has to have a knack for color tv. There's no big scarcity of color technicians, yet excellent ones are not exactly plentiful either. The name of the game, for you, is finding the right guy for whenever your color set goes bad. Be giving it some thought before that happens. Know ahead of time who you might call. The pages that follow suggest how to find him (and how not to) and what to do with him once he's agreed to take on your repair job. With color tv, repair satisfaction is a two-way street.

DON'T PUT YOURSELF IN EITHER OF THESE PICTURES. Agreed, you might save some money by taking off the back and "trying a few things" yourself. You might just as easily foul up a few things yourself. When you don't know what to expect, you're just safer to leave that back on. The best advice you can follow when you're tempted to open up the back: DON'T.

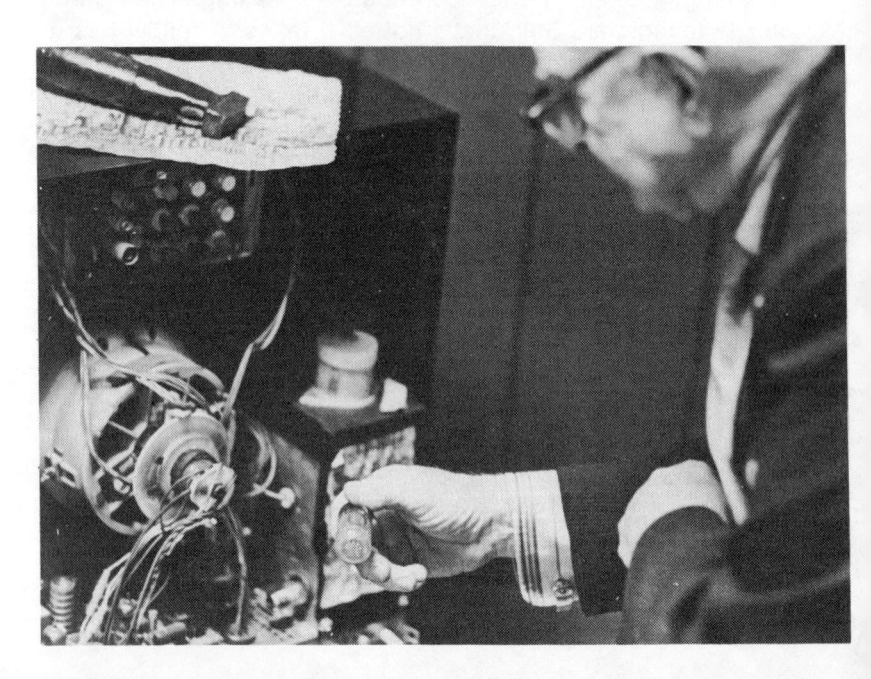

HERE'S ANOTHER SITUATION TO AVOID. The neighbor's boy may be a budding Sarnoff. He may have gone through every book the school library has and taken both courses in electronics at the vocational high school. But you may find he just can't quite handle the majority of complaints that can develop in your color television receiver. That takes time and experience, and you probably don't want to let him gain that experience on your color set.

Two-to-one, he'll be just as relieved not to get in over his head as you are to have him not. Treat the situation tactfully but firmly. It may be especially ticklish if he's the favorite nephew of Aunt So-and-So and she recommended him. Family tv "experts" can be a real hassle. Try to fend them off as smoothly as you can. But fend them off.

137

INSTEAD, DO THIS. Get on the telephone and call someone who is legitimately in the color tv repair business. As when you hunted a dealer to buy from, the Yellow Pages is one place to start.

If you know nothing about any of the shops in town, you're probably better off picking one that's not far away from you. In larger cities, shop owners frequently limit the distance they can profitably go for a flat service-call fee.

Price-shop if you want to, but nowadays that seldom saves you enough to be worth the trouble. If you do come across a really low-rate shop, you'd be well advised to pass it by; you don't get something for nothing anywhere anymore.

But there are some ways you can judge the kind of shop you expect to do business with. After you've checked with the Better Business Bureau, try some of the ideas on the next few pages.

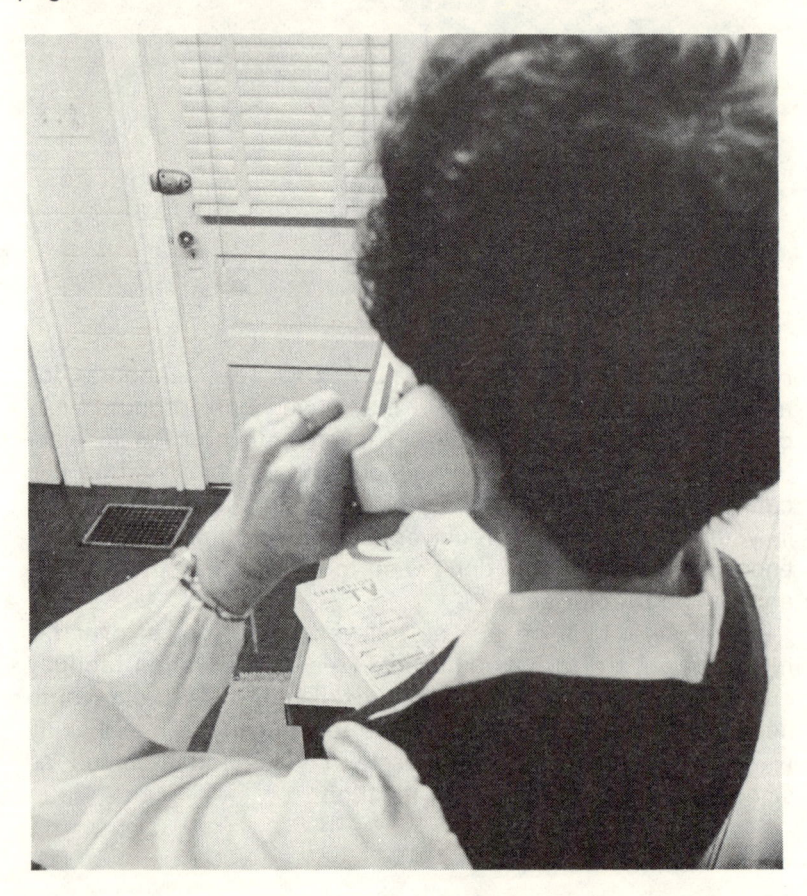

HOW THEY ANSWER THE PHONE sometimes tells you a lot about a service business of any kind. When you receive brusque treatment on the telephone, you can logically—and usually correctly—assume you'll get the same curtness in other dealings with that company.

Does whoever answered the phone mind if you ask questions about their service? Those that have nothing to hide don't mind. Their service prices should be open and discussed freely. The nature of trouble with your color set should be requested and taken down carefully. Throughout the phone conversation, you should be treated courteously, though not patronizingly. A friendly, businesslike manner usually signifies a shop that conducts servicing the same way.

IF YOU HAVE A CHANCE TO SEE THE SHOP, do. Know who you're doing business with, even before you start if possible. Know the kind of place they keep.

A television repair shop will always look cluttered if any work is going on. But it need not be dirty. Floors can be swept. A coat of paint every year or two can keep the inside and outside shaped up and presentable.

What does the service truck look like? A dilapidated truck or shop doesn't directly affect the kind of work a technician can do. But experience has proven that the guys who have genuine pride in their work won't drive around for long in a crummy-looking truck, or sit for long at a bench that's falling apart, or work indefinitely with equipment that can't do the job.

So . . . how the shop looks inside and out can mean something to your choice of a color television service technician.

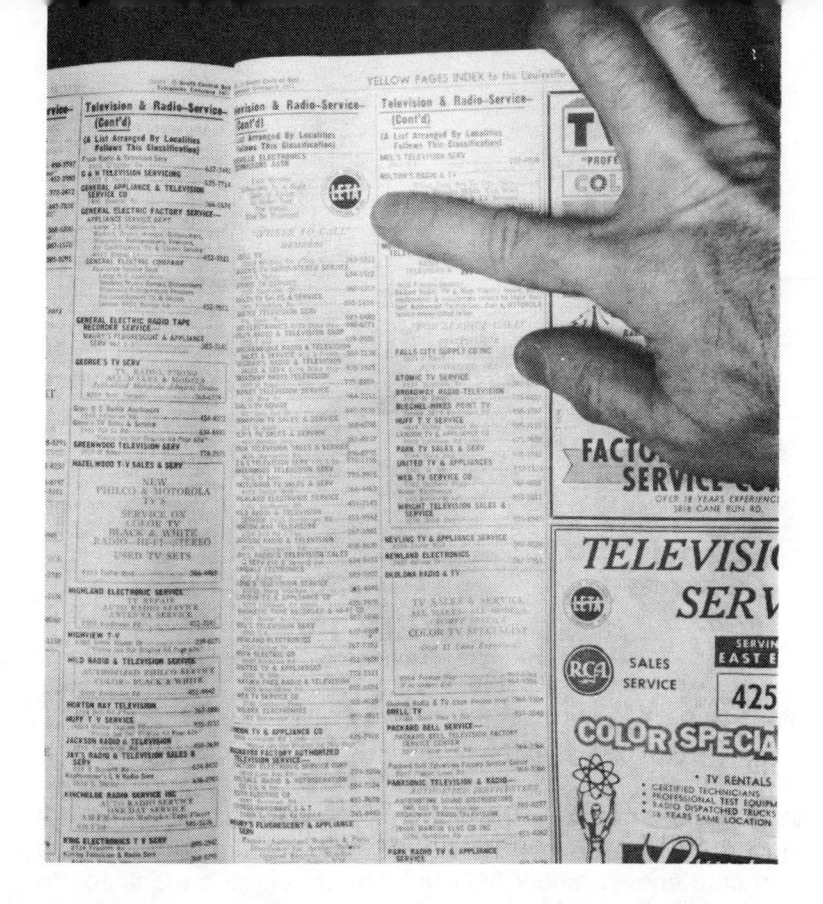

HERE'S ONE WAY TO PICK OUT A TECHNICIAN who might do you a better job than some others. In the ads for some repair shops, you'll find a small symbol identifying them with the local electronic service association. Some cities have a special grouping of associated tv shops, under "Television and Radio—Service" in the Yellow Pages. The membership emblem usually heads the grouping.

Association membership is no magic guarantee that a shop has either competent technicians or honest management. But the likelihood is greater. Some associations put teeth into their codes of ethics, tossing out members who don't measure up to acceptable practices.

So . . . when you're at a loss for how to find a technician to depend on, look for an association emblem. But then look beyond that, and apply the same criteria you'd apply to any other service shop: neatness, courtesy, equipment to do the job with, and so on.

141

SYMBOLS OF NATIONAL ASSOCIATION membership carry weight in the television repair industry, too. As with membership in local associations, you have no automatic guarantee of either ethical conduct or technical proficiency. But the odds are higher, on both counts. Service technicians and shop owners who care enough about their industry to join an association to further it seem more likely to make an extra effort to keep the public from thinking badly of their profession.

The television servicing industry has two national associations: National Electronic Associations (NEA), 1715 Expo Lane, Indianapolis, IN 46224, and National Alliance of Television and Electronic Service Associations (NATESA), 5906 S. Troy Street, Chicago, IL 60629. Both are formed of state and local associations, but have a few individual members in rural areas. Both have codes of ethics, though not enforced as strictly as some local codes of ethics are.

If you have difficulty with a member of either association, however, don't hesitate to write national headquarters. A word from higher up might trigger aid you didn't expect.

The NEA administers a Certified Electronic Technician (CET) program. To gain this recognition, the man must pass a rigid examination in television servicing. Again, it doesn't guarantee good service but it does show that he has proven he knows both theory and practice.

WHEN YOU'VE PICKED THE BEST COLOR TV TECHNICIAN you can find, what then? You should already know he's qualified and what his service charges are. You're satisfied he's competent and honest. About all you can do at this point is let him in, show him to the ailing color set, and stay out of his way. Keep the kids and the pets away while he's there servicing your color set. He doesn't need the help or the distractions.

He has a right to expect some things of you, too. If you say come in the afternoon, you owe it to him to be there or to call and postpone. You should never call more than one shop at a time. Keep that up and you may find you can't get *any* shop to send a man out when you call; word gets around.

If he fixes the set there, have the money ready to pay him. Rarely is he authorized to accept you as a credit customer. That's a job for his boss, and you should have that arranged before the man arrives. The majority of service shops work by cash-on-delivery only. Don't embarrass either of you by expecting otherwise.

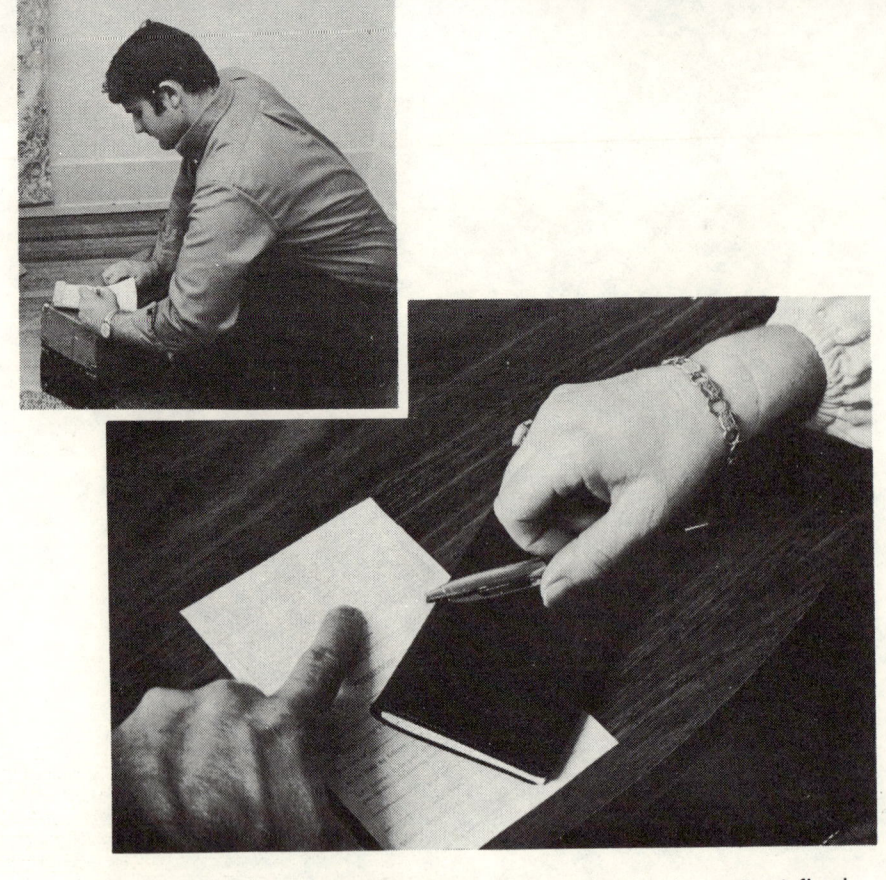

WHEN THE JOB IS COMPLETED, there are two important final points.

1. Get a specifically itemized invoice. Some states require by law that the technician leave you such an invoice. But get one, anyway. It should show the price of every part installed and show the service call and bench labor separately. You should be given back any old parts; that's also required in some states.
2. Understand fully what guarantee you have. Again, some state laws require that the service warranty be printed or written on the invoice. If it isn't, ask that it be. You have a right to be told, in writing, what you can expect the service guarantee to cover and for how long.

Legitimate servicers won't mind a bit when you ask for these important items. Don't write the check or pay the invoice until you have them.